TOXICOLOGY
AND
BIOLOGICAL MONITORING
OF METALS IN HUMANS

TOXICOLOGY AND BIOLOGICAL MONITORING OF METALS IN HUMANS

Including Feasibility and Need

Bonnie L. Carson
Harry V. Ellis III
Joy L. McCann

CRC Press
Taylor & Francis Group
Boca Raton London New York

CRC Press is an imprint of the
Taylor & Francis Group, an **informa** business

First published 1986 by CRC Press
Taylor & Francis Group
6000 Broken Sound Parkway NW, Suite 300
Boca Raton, FL 33487-2742

Reissued 2018 by CRC Press

Library of Congress Cataloging-in-Publication Data

Carson, Bonnie L.
    Toxicology and biological monitoring of metals in humans.

    Bibliography: p.
    Includes index.
    1. Metals—Toxicology. 2. Metals—Physiological effect. 3. Blood—Analysis and chemistry. I. Ellis, H. V. II. McCann, J. L. III. Title. IV. Title: Biological monitoring of metals in humans.
RA1231.M52C36   1986   615.9'253   85-23167
ISBN 0-87371-072-X

A Library of Congress record exists under LC control number: 85023167

ISBN 13: 978-1-315-89828-5 (hbk)
ISBN 13: 978-1-351-07738-5 (ebk)

Visit the Taylor & Francis Web site at http://www.taylorandfrancis.com and the CRC Press Web site at http://www.crcpress.com

PREFACE

Persons in many disciplines besides toxicology per se need information about toxic effects of substances related to exposure and how to monitor exposure. These disciplines include industrial hygiene, occupational medicine, clinical chemistry, public health, epidemiology, and environmental regulation. Anyone who has ever tried to track down information on the toxicology, exposure, and monitoring of the stable metals and metalloids knows that such information is spread among several secondary sources as well as the regulatory and primary literature. In addition, no one toxicology reference covers all the stable metals. The current volume serves a definite need in bringing toxicological, exposure, and monitoring information about the metals together in a one-stop source in a brief, uniform format.

As one of many tasks of an analytical chemistry program sponsored by the Office of Pesticides and Toxic Substances, Field Studies Branch, U.S. Environmental Protection Agency (EPA) (EPA Prime Contract No. 68-02-3938), Midwest Research Institute (MRI) prepared these summaries of the mammalian toxicology of 52 individual elements (metals and metalloids) and the lanthanides (rare earth elements). MRI is assisting the EPA in developing protocols for the monitoring of selected metals and organic compounds in the blood of the general U.S. population. The monitoring will enable evaluation of changes in the blood levels over time and any trends associated with instituted control measures. This review will help EPA select the metals to be included in the monitoring effort based on toxicity, relative exposure, and the ability of blood measurements to reflect exposure.

v

ACKNOWLEDGMENTS

The program at MRI is under the general supervision of Dr. John E. Going, Director of the Chemical Sciences Department at MRI. Analytical chemists Dr. Lloyd M. Petrie and Mr. Ed Lustgarten evaluated the methodological quality of the primary analytical papers mentioned in the summaries. We acknowledge the support and patience of Ms. Janet Remmers, Task Manager, and Drs. Frederick Kutz and Joseph J. Breen, former and current project officers, respectively, U.S. Environmental Protection Agency, Office of Pesticides and Toxic Substances, Field Studies Branch, and the reviewers of the draft report.

Bonnie L. Carson is a Senior Chemical Information Scientist at Midwest Research Institute (MRI), Kansas City, Missouri, where she specializes in compiling reports on the environmental chemistry and health effects of toxic substances from information published in the world literature. She was the major author and editor of six of the seven reports to the National Institute of Environmental Health Sciences that were eventually published as the six-volume series *Trace Metals in the Environment.* Other major projects she has led at MRI include a series of reports detailing the inhalation health effects of chemicals emitted from automobiles equipped with catalytic converters for the U.S. Environmental Protection Agency's Emission Control Technology Division and technical data entry for EPA's Oil and Hazardous Materials Technical Assistance Data System (OHM-TADS). Before coming to MRI in 1973, Ms. Carson served brief stints as an organic chemistry laboratory instructor, an abstractor at Chemical Abstracts Service, and a free-lance Russian translator. Among her professional affiliations are the American Chemical Society, the American Society for Information Scientists, the American Translators Association, the New York Academy of Sciences, and the Society for Technical Communication. She is on the editorial board of *Science of the Total Environment* and is listed in *Who's Who in the Midwest* and *American Men and Women of Science.* Ms. Carson received her BA in chemistry *summa cum laude* from the University of New Hampshire in 1963 and her MS in organic chemistry from Oregon State University in 1966.

Dr. Harry V. Ellis III, Toxicologist, PRC Engineering, Chicago, Illinois, worked with Ms. Carson and Ms. McCann for several years on the staff of and as consultant to MRI. Besides reviewing the health effects of environmental chemicals for reports coauthored with Ms. Carson, Ms. McCann, and other MRI staff, Dr. Ellis served as principal investigator and/or study director on multidisciplinary experimental toxicology studies on various environmental chemicals and chemotherapeutic agents. Dr. Ellis is a major in the U.S. Army Reserve, serving as an Environmental Science Officer. From 1971 to 1975, Dr. Ellis was research pharmacologist, captain, at the Walter Reed Army Institute of Research, Washington, D.C. Dr. Ellis is affiliated with the American Chemical Society, the New York Academy of Sciences, the American Society for Testing and Materials, the American Defense Preparedness Association, and the Association of the United States Army. Dr. Ellis received a BS degree in chemistry from Massachusetts Institute of Technology in 1965 and a PhD in pharmacology from Emory University in 1971.

Joy L. McCann, Assistant Environmental Scientist at Midwest Research Institute, specializes in the compilation of literature and solicited information on environmental and health aspects of toxic substances. She is a coauthor of the chapter on physiological effects in *Trace Metals in the Environment: Volume 6, Cobalt*, and of the reports to EPA on the inhalation health effects of sulfuric acid, hydrogen sulfide, formaldehyde, and methanol. She has been a major contributor to the nationally available online database OHM-TADS for several years. Ms. McCann joined MRI in 1976 following graduation from Central Missouri State University with a BS in agriculture. She has served as a general assistant to the editor of *Acres U.S.A.*, a monthly newspaper covering ecoagriculture in North America.

TABLE OF CONTENTS

---

*   See "Yttrium and the Lanthanides (Rare Earths)."
** See "Niobium."

Page

* See "Yttrium and the Lanthanides (Rare Earths)."

---

* See "Yttrium and the Lanthanides (Rare Earths)."

---

* See "Yttrium and the Lanthanides (Rare Earths)."

I. INTRODUCTION

A. Objective

The Office of Toxic Substances (OTS) is planning a National Blood Network similar to the National Human Adipose Tissue Survey (NHATS), currently administered by the Exposure Evaluation Division of OTS. Selected metals and organics will be monitored in the blood of the general U.S. population so that changes in the levels over time can be assessed.

Midwest Research Institute (MRI) compiled brief summaries on the health effects from metal exposures for use in identifying specific metals to be included in the analytical chemistry protocols for the OTS's human monitoring initiative. The health effects compilation will also aid in interpretation and evaluation of the results from the chemical analyses.

B. Methods

Elements chosen for review included all metals for which some toxicity other than radiation effects might be expected, plus selenium, tellurium, and the metalloid arsenic.

Authoritative secondary toxicology references were the main sources of information. The major sources for most elements were the chapter on metals from the third edition of Patty's Industrial Hygiene and Toxicology and the Toxicology of Metals, Volume II, an extensive report prepared for EPA's Health Effects Research Laboratory in 1977. Other secondary sources included the NIOSH criteria documents for occupational exposure, EPA's ambient water quality criteria documents for the priority pollutant metals, National Academy of Sciences reviews, and the books based on MRI reports to the National Institute of Environmental Health Sciences (NIEHS) for the project "Assessment of Environmental Exposure to Heavy Metals."

In addition, recent primary references were identified from a computer search of toxic substances in human blood. Analytical chemists rated the quality of the analytical methodology. Any information in the individual metal profiles linking exposure and/or toxicity to blood concentrations includes MRI's qualitative rating.

C. Organization

Following the introduction is a summary and 53 profiles. One profile was prepared for each of the elements except the chemically related lanthanides, which are mentioned in one profile, "Yttrium and the Lanthanides (Rare Earths)."

II. SUMMARY

Based on their known or suspected adverse health effects, blood monitoring of the general population would be useful for antimony, arsenic, beryllium, cadmium, chromium, cobalt, lead, lithium, mercury, molybdenum, nickel, sodium, strontium, or thallium. Selenium might be monitored in consideration of its reported cancer-protective aspect.

Blood values correlate reasonably well with degree of exposure for cadmium (only after ~ 2 yr; note that smokers may have higher levels than nonsmokers), lead, mercury, nickel (some authors do not recommend blood monitoring), and selenium (but concentrations may not be related to toxicity). Blood values increase upon exposure; but the increases are not dose-related or apparently have not been quantitatively related to exposure for beryllium, copper, lithium, manganese, vanadium, and zinc.

Some authors have definitely stated that blood monitoring is not recommended for chromium, nickel, and thallium. Palladium and platinum could not be detected even in refinery workers' blood.

Urine analysis is feasible or recommended for monitoring exposure to aluminum, arsenic, chromium, cobalt, nickel (plasma better), selenium, thallium, uranium, and vanadium. Neither blood nor urine values correlate well for antimony exposure.

Blood monitoring for exposure to certain elements may not be useful because of rapid clearance (gallium, germanium, molybdenum, rubidium, silver, thallium, uranium, and lanthanum and the other rare earth metals), because of very slow clearance (beryllium), or because the element's con- centration is under homeostatic control (calcium, magnesium, potassium, perhaps selenium once levels attain 200 to 240 µg/L, sodium, and zinc). There is significant sex difference or diurnal, day-to-day, and/or week- to-week variation in blood values for copper, iron, and tin.

Table 1 summarizes the known relations, if any, among exposure level, toxic effects, and blood or urine concentrations for all the elements considered.

III.   METAL PROFILES (Mammalian Toxicity Summaries)

Following are the 53 profiles. One profile was prepared for each of the elements except the chemically related lanthanides, which are men- tioned in one profile, "Yttrium and the Lanthanides (Rare Earths)." Each of the lanthanide elements is mentioned in the table of contents and cross referenced to that profile. The extensive material available was evaluated and the essential information extracted was presented in a standard format to facilitate comparisons and ready reference.

Each profile is self-contained and formatted like a short report. The introduction to each profile includes information on the occurrence and production, uses, and chemistry of the element. Subsequent sections are on exposure (and any exposure limits), toxicokinetics, effects, and references.

TABLE 1.   HUMAN LOW-LEVEL EXPOSURE TO METALS AND BIOLOGICAL MONITORING

Element	Exposure level	Biological effect(s)	Concentration in biological media[a] (μg/L)	Comment
Aluminum	~45 mg/day[b]	-	<10 (s)	
	Occupational (TLV[c] is 2 mg/m³ for sol. salts and 10 mg/m³ for metal and oxide)	-		Urine values increase with industrial exposure, but there is not much interest in biol. monitoring.
	Large oral doses of Al(OH)₃	Hypophosphatemia	No data[d]	
Antimony	~50 μg/day[b]	-	3 (w) ~0.5 to 2.6 (u)	
	Occupational (PEL is 0.5 mg Sb/m³)	Heart disease, liver damage, G.I. ulcers higher prevalence in workers	5,000 to 200,000 (w) 5,000 to 182,000 (u)	Neither blood nor urine values correlate well with degree of exposure.
	~3 mg/m³	Heart disease, liver damage, G.I. ulcers higher prevalence in workers.	800 to 9,600 (u)	A biol. threshold has not been proposed.
Arsenic	~1 mg/day[b]	-	<100 (u)	
	>50 μg/L (U.S. drinking water standard)	Cancer, cardiovascular disease, and neurol. effects	No data	
	6 to 123 μg/L drinking water 393 μg/L	No data	0.29 to 0.51 (w)	Blood values did not correlate well with exposure levels. Hair and urine were much more useful.
		No data	1.33 ± 1.18 (w)	
	Occupational (PEL[e] is 0.01 mg/m³ for inorganic As)	Chronic occupational effects include skin lesions, peripheral neuritis, G.I. disorders, lung and skin cancer.		Occupational exposure is monitored by urine analysis.
Barium	~750 μg/day[b]	-	80 to 400 (w) (most in plasma)	Ba disappears from the blood 24 h after dosing.

TABLE 1. (continued)

Element	Exposure level	Biological effect(s)	Concentration in biological media[a] (μg/L)	Comment
Barium (cont'd)	Occupational (PEL is 0.5 g/m³ for sol. compds.)	Benign pneumoconiosis	No data	
Beryllium	~12 or 100 μg/day[b]	-	<0.1 to 3.8 (w)	Be in blood and urine has been only qualitatively related to exposure. Be in urine may be detected for years after removal from exposure.
	Occupational (PEL is 0.002 mg/m³)	Immune system interference. Lung cancer proved in animals but not humans.	No data found for workers or other exposed persons.	
Bismuth	~20 μg/day[b]	-	No data	
	Occupational (PEL is 20 mg Bi_2Te_3/m³)	Tellurium breath but no adverse effects	No data found for workers or other exposed persons.	
Boron	~1 mg/day[b]	-	No data for normal or occupational blood levels.	
Cadmium	~150 μg/day[b]		<10 (w) (most bound to rbc) 0.05 μg/g (a)	Smokers have about twice the body burden nonsmokers.
	Chronic occupational (PEL is 0.2 mg dust/m³)	No adverse effects at corresponding blood level	10 (w)	
	0.7 mg cadmium-containing dusts per m³	Proteinuria (2 workers). Breathing difficulty and renal tubular damage (1 worker)	18 to 24 (w) 35 (w)	Some authors state that blood concns. correlate reasonably well with worker exposure only after ~2 yr. Others find blood and urine levels correlate with acute exposure but cannot predict possible kidney damage.
	1.5 to 2 mg dust/m³	Breathing difficulty and/or renal damage.	44 to 73 (w)	
Calcium	~1,100 mg/day[b]	Fatal Tetany Convulsions	100,000 (p) >160,000 (s) 70,000 (p) 35,000 to 50,000 (p)	Ca in blood is under homeostatic control. Hypercalcemia occurs in certain disease states but is not caused by elevated exposure.

TOXICOLOGY AND BIOLOGICAL MONITORING   5

TABLE 1. (continued)

Element	Exposure level	Biological effect(s)	Concentration in biological media[a] (µg/L)	Comment
Cesium	~10 µg/day[b]	No chronic toxicity reported in humans or animals.	No data	Cs concentrates in rbc.
Chromium	~150 µg/day[b]	—	0.160 (p or s)	Blood Cr is not of value in monitoring exposure; urine is. Cr in normal blood is distributed equally in rbc and plasma, but occupational exposure increases Cr mainly in rbc.
	(TLV is 0.05 mg/m³ for water-sol. Cr(VI) compds.)	Chronic inhalation exposure to Cr(VI) causes respiratory tract lesions, lung cancer, G.I. disturbances, and other problems.	30 µg Cr/g (u) (>6X greater than normal)	
Cobalt	~50 or 300 µg/day[b]	—	≤238 (w), ≤43 (p), ≤660 (s), ≤206 (rbc), 100 to 750 (u)	Highly variable blood values reported, perhaps due to contamination.
	≥12 mg/day (cobalt treatment for anemia for up to 8 months)	Toxic symptoms (e.g., nerve deafness and goiter)	No data	
	Occupational (PEL is 0.1 mg/m³ for metal, dust, and fume)	Lung disease (e.g., fibrotic changes, asthma)	No data	Urinary Co correlates well with occupational exposure.
	1 mg Co/L beer	Alcohol content plus cobalt caused fatal congestive heart failure.	No data	
Copper	3.5 mg/day[b]	—	160 to 3,480 (w), 700 to 1,400 (s) (men), 850 to 1,550 (s) (women), 930 (rbc), 0.24 µg/g (a)	Abnormally high values in this 19-city survey were from people living in Cu-smelting vicinities. There is diurnal variation and sex differences. Day-to-day variations higher in females.
	Occupational (PEL is 0.1 mg Cu/m³ for fumes)	Metal fume fever	>1,600	No correlation with degree of exposure found.

TABLE 1. (continued)

Element	Exposure level	Biological effect(s)	Concentration in biological media[a] (µg/L)	Comment
Copper (cont'd)	0 6 - 1 mg/Cu/m³	Mild anemia	1,077 ± 40 (p) (insignificant difference); 28 0 ± 2 6 (rbc) (significantly different from control value of 22.5 ± 2 4).	
Gallium	No data	No data	No data	Ga remains in the blood only a short time
Germanium	No data	No data	No data	Ge is cleared from the blood stream within a few hours after dosing.
Gold	No data on normal exposure	-	<2.5 (w) ≤20 (u) 0 011 µg/g (a)	
	2 53 g in 4 yr	Toxic symptoms (dermatitis and nephrotic syndrome expected at this level)	9,500 (h) 9,900 (u)	Patients who develop toxicity do not show statistically significant differences in serum Auconcens.
Hafnium	~4 µg/day[b]; 0.5 mg/m³ is OSHA PEL	No chronic effects information for humans; borderline liver toxicity seen in rats	No data on normal or occupational blood or urine values	Clearance is slow.
Indium	≤8 µg/day[b]	-	No reliable data	In is bound to rbc.
Iridium	No data	No chronic ill effects attributed to In	No data	
Iron	~16 mg/day[b]	No data	1,290 (p) (men) 24 µg/g (a)	Normal serum values vary considerably during the day and from day-to-day or week-to-week.
	Occupational (TLV is 1 mg fe/m³ for sol. fe salts)	Iron pneumoconiosis	~1,600 (s)	

TABLE 1. (continued)

Element	Exposure level	Biological effect(s)	Concentration in biological media[a] (μg/L)	Comment
Lead	450 μg/day[b]	-	<350 (w) (50% ≤ 200) (300 is current max. acceptable level) 0.04 μg/g (a)	Pb in single blood sample is reliable exposure monitor in steady-state. Most Pb in blood is in rbc.
	Occupational (PEL is 0.05 mg/m³ for Pb and inorg. Pb compds.)	Subclinical CNS effects. Chronic poisoning symptoms include G.I. disturbances, weakness in extensor muscles, tremors, and encephalophathy	>500 (w) >700 (w)	
Lithium	~2 mg/day[b]	- Chronic toxicity includes effects on G.I. tract, CNS, and kidney. Li in drinking water significantly correlated with arteriosclerotic heart disease.	~17 (p)	Plasma concns. increase with increased exposure, but increases not dose-related.
Magnesium	340 mg/day[b] or higher	-	1.807 ± 0.132 mEq/L (s) (~44,000 μg/L) 20 μg/g (a)	Blood levels are not a good indicator of body stores or long-term exposure. Homeostatic regulation of G.I. absorption and renal excretion maintain const. serum levels.
	500 to 1,000 mg/L drinking water	Cathartic effect on initial exposure. No serious chronic effects from excess Mg reported.		
Manganese	3.7 mg/day[b]	-	0.587 ± 0.183 (s) 0.015 μg/g (rbc) <3 (u) 0.03 μg/g (a)	Mn concentrations in rbc elevated in arthritic persons.
	Occupational (TLV is 1 mg/m³ as fumes)	Toxic CNS symptoms	>40 to 50 (u)	A significant relation between exposure and blood concns. has not been found in workers. Correlation with chronic effects may not be possible because of differences in individual susceptibility.
Mercury	15 or <10 μg/day[b]	-	<20 (w) 0.30 μg/g (a)	About twice as much inorganic mercury is in rbc as in plasma, but 90% of methylmercury is in rbc.

TABLE 1. (continued)

Element	Exposure level	Biological effect(s)	Concentration in biological media[a] (µg/L)	Comment
Mercury (cont'd)	Occupational <0.1 mg/m³	No significant differences from controls	19.9 (w)	
	Occupational (PEL is 0.01 mg/m³ for alkylmercury, 0.05 for Hg vapor, and 0.1 for other compounds)	Some clinical evidence of toxicity (e.g., protein-uria)	~30 to 35 (w) (Hg vapor at PEL) >35 (w)	
	Chronic methylmercury exposure in fish	Paresthesia and sensory disturbances are earliest signs.	>200 (w) 50 µg/g (h)	Each microgram of methylmercury ingested contributes 0.8 µg Hg/kg blood.
	Chronic methylmercury exposure in fish	No signs attributable to methylmercury	82 (w) (9.9 in population eating 20% as much fish)	
	Single meal of methyl-mercury-containing fish (18 to 22 µg Hg/kg bw)	-	40 to 60 (w)	Peak 4.7 to 14 h after ingestion.
Molybdenum	~300 µg/day[b] 150 to 500 µg/day is RDA	-	5.0 to 157.3 (w) (detected in 48 of 229 samples)	No clears from the blood within 24 h. No concns. decrease in anemia. Detns. in serum are subject to gross errors due to sample contamination.
	540 µg/day in diet	Increased Cu in urine	No data	
	10 to 15 mg/day (Armenia)	Gout-like disease	No data	
	Occupational (PEL is 5 mg/m³ for sol. compds.)	-	No data	
Nickel	~400 µg/day	-	2 to 3 (s) 2 to 4 (p) 0.035 µg/g (a)	Persons exposed to toxic doses have higher dose-related Ni levels in serum. Sample contamination often leads to gross anal. errors.

TABLE 1. (continued)

Element	Exposure level	Biological effect(s)	Concentration in biological media[a] (µg/L)	Comment
Nickel (cont'd)	Occupational (PEL is 1 mg/m³ but 0.015 mg/m³ recommended)	No adverse effect at <10 µg/L plasma	<10 (p)	24-h urine samples may correlate well with worker exposure; morning and after-work plasma values correlate better than single urine values. Some authors do not recommend monitoring blood levels.
	Occupational (sol. Ni compds.)	The only effect of chronic Ni exposure studied in humans is respiratory tract cancers.	7.4 (p) (4.2 for controls)	
Niobium (Columbium)	620 µg/day[b] No PEL or TLV	No reports of chronic (or even acute) toxicity in humans	530 to 740 µg/L (s) 4,190 to 6,400 µg/L (p)	Values based on limited data from Schroeder and Balassa (1965). Their analyses for many other elements are often one or two orders of magnitude higher than those reported for the same element by other researchers.
Osmium	≥0.1 mg OsO₄/m³ Occupational (PEL is 0.002 mg OsO₄/m³)	Eye irritation	No data for normal or occupational values.	
Palladium	–	No information on effects in humans from chronic exposure.	No data for normal or occupational values.	Even palladium refinery workers have no detectable palladium in their blood (<4 µg/L).
Platinum	>2 to 20 µg/m³ (PEL is 2 µg/m³ for sol. Pt salts)	Platinosis—an asthma-like condition with dermatitis.	No data for normal or occupational values.	Workers' urinary Pt levels can be detected, but concns. are below detection limits.
Potassium	3.3 mg/day[b] No PEL or TLV applicable	No chronic effects documented	140,000 to 215,000 (s) 320 µg/g (a)	Most K^+ is in rbc. K^+ in blood is under homeostatic control, but hemolysis, transport from plasma to rbc during storage, and other factors can affect accuracy of anal. detn.
	40 to 90 mEq/day	Deaths due to sudden cardiac arrest in 1 of every 400 hospital patients receiving such doses (patients had unsuspected renal insufficiency)	No data	

TABLE 1. (continued)

Element	Exposure level	Biological effect(s)	Concentration in biological media[a] (μg/L)	Comment
Rhenium	No information	No information on chronic (or acute) effects in humans.	No data	
Rhodium	PEL is 1 μg/m³ for sol. Rh salts	No information on effects from chronic exposure.	No data	
Rubidium	2.2 mg/day[b]	No information on effects in humans from chronic exposure.		Rb is quickly distributed by the blood to other tissues with some being retained by the rbc. Normal urinary losses are estimated at 1.9 mg/day.
Ruthenium	Soviet researchers recommended max. permissable concns. of 1 mg RuO₂/m³ and 0.1 mg RuOHCl₂/m³	No information on human adverse effects from chronic exposure.	No data	
Scandium	0.2 μg/day[b]	No human information	<0.03 (w)	
Selenium	150 μg/day[b]	-	157 to 265 (w) (U.S.)	Se in blood and urine reflect recent exposure. Blood concns. may not reliably predict clinical toxicity because part of body stores may not be under homeostatic control. Even experienced labs report erroneous plasma Se concns.
	Keshan disease area, 8 μg Se/day	Selenium deficiency syndrome	7 to 79 (u)	
	Endemic selenosis area in China, 4,990 μg Se/day	-	3,180 (w)	
	26 to 1,800 μg Se/L drinking water	Adverse effects from chronic exposure not proved in humans	133 to 248	There may be homeostatic control of blood Se concns. once blood levels attain 200 to 240 μg Se/L.
Silver	70 μg/day[b]	-	24 (w) N.D. to 14 (s)	Ag is assocd. with plasma proteins. Blood clearance is rapid.
	≥1 g total dose	Chronic exposure to Ag by all routes causes gray skin and eye discoloration called argyria and argyrosis, respectively.	Found in urine only in poisoning cases.	

TABLE 1. (continued)

Element	Exposure level	Biological effect(s)	Concentration in biological media[a] (µg/L)	Comment
Sodium	2,300 to 6,900 mg/day	Long-term excessive Na intake is one factor assocd. with human hypertension.	2,900,000 to 3,560,000 (s); 510 µg/g (a)	Most Na^+ is in the plasma. Levels are under homeostatic control.
Strontium	~1 to 2 mg/day[b]	Excessive chronic intake might cause disturbance in mineral metabolism.	29 (16 to 43) (p); 25 µg/g (a)	Most blood Sr is in the plasma.
Tantalum	Background (no data)	-	16 (w); 5; 240 (u)	Based on one blood and two urine samples; anal. by spark source mass spectrometry.
	Occupational (PEL is 5 mg/m³)	Early lung fibrosis seen in Russian workers exposed to Ta and Nb.	No data	
Tellurium	~0.6 mg/day	-	0.001 to 0.07 (p); based on 0.5 to 7% of urine concn. (0.2 to 1.0) (u)	
	0.4 to 0.88 g K tellurite for 7 days	Drowsiness, anorexia, nausea, cardiac oppression, breath odor lasting for 7 wk	No data	
	0.1 to 1.0 mg Te/m³	Me_2Te breath odor, drowsiness, possibly other symptoms	No data	
Thallium	~2 µg/day[b]	-	1.1 µg Tl/g creatinine (u); ≤0.8 (u)	Blood is not a reliable indicator of Tl exposure. Even in a poisoning victim, practically all of the Tl clears from the blood within 2 h. Urine is a better indicator of exposure. Hair also might be used.
	≥7 mg/kg for Tl(I) compds	Acute poisoning	≥50 (u)	
	Chronic environmental contamination Occupational (PEL is 0.1 mg/m³)	Sleep disorders and neurological symptoms. Alopecia may be a sign of acute or chronic exposure.	≥20 (u)	

TABLE 1. (continued)

Element	Exposure level	Biological effect(s)	Concentration in biological media[a] (μg/L)	Comment
Thorium	~3 μg 1 day[b] Maximum permissable concn. is 10⁻⁶ μCi/mL air.	Radiotoxicity is the chief concern of long-term exposure to excessive Th	No data	
Tin	4 mg/day[b]	—	120 to 140 (w) 20 (p)	Almost all Sn is in rbc. Concns. among some populations varied considerably from day-to-ay and week-to-week. Variations due to diet.
	Occupational (OSHA PEL for inorganic Sn is 2 mg Sn/m³; for organotins, NIOSH recommended 0.1 mg Sn/m³).	Benign pneumoconiosis from Sn oxide dust. No other information on toxicity from chronic human exposure.	No data	
Titanium	~850 μg/day	No chronic effects have been attributed to titanium.	20 to 70 (w) 10.2 (u) 0.031 μg/g (a)	
	Occupational (coal miners)	TiO$_2$ and Ti are practically inert.	~2,300 and 3,000 (w) 110 and 490 (u)	
Tungsten	8 to 13 μg/day[b]	—	No data on normal levels.	
	Occupational (TLV is 1 or 5 mg W/m³ for sol. or insol. W compds; Co or Ni guidelines may take precedence)	Hard metal disease (cough, dyspnea, and wheezing with minor radiol. abnormalities) is primarily due to Co content in cemented WC	No data	
	Occupational 0.75 to 6.1 mg W/m³ and 0.6 to 3.2 mg Co/m³	Symptoms of 88/178 workers included dyspnea, coughing, headache, dizziness, G.I. disturbances, impaired sense of smell	8,000 to 11,000 (w) (45 workers; not detected in 11); 600 to 1,100 (u) (40 workers)	
Uranium	~1.9 μg/day	—	(20% is in rbc, 79% in plasma) 0.03 to 0.3 (u)	Clearance is rapid from the blood stream.

TABLE 1. (continued)

Element	Exposure level	Biological effect(s)	Concentration in biological media[a] (µg/L)	Comment
Uranium (cont'd)	Occupational (PEL is 0.05 mg/m³ for sol. compds.)	Respiratory disease and cancer Kidney damage	No data >250 (u) (post shift)	
Vanadium	~2 mg/day	-	<1 (w) 0.94 ± 0.65 to 2.83 ± 2.20 (u)	
	0.01 to 0.04 mg V/m³	No hematol. abnormalities	5.1 ± 3.2 (given as 0.22 ± 0.14 µmol/L)	Urine values correlated with exposure but blood values did not.
	Occupational (cleaning of oil-burning boiler)	Chronic exposure to V in air may cause respiratory difficulties.	0.83 ± 0.29 to 5.45 ± 3.61 (w)	
Yttrium and the Lanthanides (Rare Earths)	No information	No human information	No data in secondary references	La and other rare earths are rapidly cleared from the blood.
Zinc	~13 mg/day[b]		1,000 (s) ~5,000 (w) ~10,000 (rbc)	Blood levels are homeostatically controlled. Serum Zn is lower in women taking oral contraceptives, in pregnant women, and in persons under certain stresses.
	Occupational (TLV for ZnO fume is 5 mg/m³)	Workers may experience metal fume fever, G.I. disturbances, and clinically latent liver dysfunction.	No data on blood	NIOSH does not recommend monitoring blood. Contamination may cause anal. errors, although serum values are generally reliable.
	135 mg/day for 18 wk to promote wound healing	Excessive Zn intakes may aggravate marginal Cu deficiency	1,570 (s) at 6 wk if initial value 950 (s), but no increases seen if initial value >1,100 (s)	
Zirconium	4.2 mg/day[b]	-	10 to 20 (w) 110 to 680 (s) 6,180 (rbc) 1 (p) (mineral part) 2.94 to 42.27 (a)	Adipose data from one report from Schroeder's lab.

TABLE 1. (continued)

Element	Exposure level	Biological effect(s)	Concentration in biological media[a] (μg/L)	Comment
Zirconium (cont'd)	Occupational (PEL is 5 mg Zr/m³)	Lung changes may be due to other constituents in the dusts.	No data	

[a] w = whole blood; s = serum; p = plasma; rbc = red blood cells (erythrocytes); u = urine; h = hair; a = adipose.

[b] Estimated daily intake by 70-kg reference man or analogous estimate.

[c] PEL = Occupational Safety and Health Administration's permissible exposure limit (standard). TLV = American Conference of Governmental Industrial Hygienists' recommended threshold limit value. Both are time-weighted averages for an 8-h day and 40-h work week. When both PEL and TLV exist for a metal or its compounds, the PEL is usually given.

[d] No data means no information was found in the secondary or primary references used. Data very probably do exist in the primary literature for many of these elements.

[e] CNS = Central Nervous System.

[f] G.I. = gastrointestinal.

The toxicokinetics section includes information on absorption; distribution (general, blood, and, if available, adipose tissue subsections); and excretion. Any information linking exposure to concentrations in blood, urine, or other tissue (usually hair) is discussed in this section. The effects section is divided into subsections on acute toxicity, chronic toxicity, biochemistry (the roles of essential elements and their recommended daily intakes are discussed in this subsection), specific organs and systems, teratogenicity, mutagenicity, and carcinogenicity.

In some profiles, and in some exposure information for most profiles, both the primary source that was cited by the secondary source and the secondary source itself were cited: for example, "McNealey et al. 1972; cited by NAS 1977." Time constraints did not permit more extensive use of this double citation method. It should be noted that even the secondary references cited are by no means exhaustive. In the profiles where "no information" or "no data" are entered, the statement more often than not merely means that the secondary sources consulted did not consider the subject.

ALUMINUM

MAMMALIAN TOXICITY SUMMARY

I. INTRODUCTION

A. Occurrence and Production

Aluminum is a common constituent of the earth's crust, usually combined with oxygen, fluorine, and silicon. Most aluminum is produced by electrolyzing bauxite in molten cryolite ($3NaF \cdot AlF_3$). Most bauxite consumed in the United States is imported (Browning 1969).

B. Uses

Uses for aluminum metal include building construction, aluminum paint, electrical uses such as overhead distribution lines, consumer durables including automobile highway signs, packaging, and containers. Bentonite and clays, natural aluminum minerals, are used in purification of water and sugar. Na_3AlO_3 and $NaAl(SO_4)_2$ (alum) may be used as coagulants in water purification. They are also used in food processing and baking. Anhydrous $AlCl_3$ is the classic Friedel-Crafts alkylating catalyst; hydrated $AlCl_3$ is an astringent used in cosmetics. Corundum (α-Al_2O_3) is used in abrasives and refractories. Alumina is used in catalysts and catalyst carriers, adsorbents, pigments, and fillers (Stokinger 1981).

C. Chemical and Physical Properties

Aluminum, atomic No. 13, density 2.7, melts at 660°C and boils at 2327°C. Its oxidation state in compounds is +3. $AlCl_3$ is slightly water soluble. Water-insoluble $Al(OH)_3$ exhibits amphoteric behavior, dissolving in strong acids or bases (Merck Index 1983).

II. EXPOSURE AND EXPOSURE LIMITS

A. Oral

Aluminum compounds are used therapeutically to prevent hyperphosphatemia in renal disease, in antacid preparations, and as an antidote. Unprocessed food usually contains < 10 mg Al/kg although some vegetables and fruits may contain up to 150 mg Al/kg. Aluminum compounds used in storing and processing food (e.g., baking powder, cooking vessels, metal foil) increase the aluminum content. Total daily intake may be $\sim$ 80 mg (Norseth 1977) A 1962-1967 survey found Al in 47.80% of the 380 drinking waters examined at 3 to 1,600 µg/L. The mean was 179.1 µg/L (NAS 1977).

B. Inhalation

Inhalation of aluminum compounds has been used in the prevention of silicosis. Nonurban air usually contains < 0.5 µg Al/m³, but urban air contains up to $\sim$ 10 µg/m³ (Norseth 1977).

16

The threshold limit value (TLV) as a time-weighted average in workroom air for an 8-h day is 10 mg/m^3 for aluminum metal and oxide; the short-term exposure limit is 20 mg/m^3/15 min. The TLV is 2 mg/m^3 for aluminum soluble salts and alkyls, 5 mg/m^3 for aluminum pyro powders and fumes (ACGIH 1983).

C. Dermal

Topical aluminum preparations are used as deodorants (Norseth 1977).

D. Total Body Burden and Balance Information

Snyder et al. (1975) give the following balance information for 70-kg reference man:

Intake, mg/day		Losses, mg/day	
Food and Fluids	45	Urine	0.10
Airborne	0.10	Feces	43
		Sweat	1
		Hair	0.0006

III. TOXICOKINETICS

A. Absorption

Uremic patients treated with oral doses of aluminum hydroxide (to bind phosphate) absorb about 15%. No information was found on absorption of other aluminum compounds by humans. Twice the normal content of Al (as the chloride or sulfate) in the food of rats did not increase the Al concentration in urine or tissues, but 15 times the normal amount did.

No data exist on absorption from the respiratory tract (Norseth 1977)

B. Distribution

1. General

Although found in all human tissues, lungs concentrate Al the most (200 to 300 g/kg), presumably from inhaled air. Increased Al in food of rats increased Al in liver, brain, testes, and blood but not the lungs. Brain accumulates the highest amounts from parenteral administration. Certain disease states influence the concentration of Al in organs, e.g., Alzheimer's disease (Norseth 1977).

2. Blood

Normal values for aluminum range from 0.037 mg/L in serum, by flameless atomic absorption spectrometry, to 6.24 mg/L for whole blood, determined by emission spectroscopy (Norseth 1977). Lauwerys (1983) cites a range of 6 to 37 µg/L mean normal serum values reported in the literature.

Versieck (1984) noted that the problem of sample contamination is apparent in many reported literature values for aluminum in serum. The true value is probably < 10 µg/L. The $^{31}P(n,\alpha)^{28}Al$ reaction in neutron activation analysis gives rise to additional aluminum concentrations of 95 µg/L.

C. Excretion

Absorbed aluminum is excreted in the urine and probably in the bile but little aluminum is absorbed from the gastrointestinal tract and little appears in the urine even after ingesting large amounts. It is eliminated as insoluble aluminum phosphates in the feces (Browning 1969; Norseth 1977).

Because of the low toxicity of aluminum in industry, there is not much interest in its biological monitoring. Urine values increase with industrial exposure, but the quantitative relations have not been determined (Lauwerys 1983).

IV. EFFECTS

A. Acute and Other Short-Term Exposures

Parenteral and oral dosing of rats with aluminum hydroxide, chloride and sulfate caused lethargy, anorexia, and death (Norseth 1977).

B. Chronic Exposure

Large oral doses of aluminum in humans tie up dietary phosphate rerouting phosphate excretion from the urine to the feces. This may lead to phosphorus depletion in the long term (Browning 1969).

Long-term feeding of rats with $AlCl_3$ at 100 to 200 mg/kg retarded growth and disturbed phosphate and carbohydrate metabolism. Small amounts of aluminum-containing food additives (1 to 2% in the diet) stimulated growth, but larger concentrations led to the metabolic disturbances.

The encephalopathy syndrome seen in uremic patients maintained on dialysis and aluminum hydroxide may be due to Al poisoning. Hypophosphatemia occurs in these patients and may occur in persons using aluminum hydroxide as an antacid (Norseth 1977). See also part IV.D.1.

C. Biochemistry

1. Effects on Enzymes

No information.

2. Metabolism

No essential biological role has been established for Al, but experiments are hampered by failure to produce Al-free diets.

3. Antagonisms and Synergisms

Oral doses of Al may induce phosphorus depletion syndrome and deplete red cell ATP (adenosine triphosphate). All compounds form insoluble phosphates in the gastrointestinal tract. Phosphorus metabolism of other tissues is also disrupted, but the mechanisms are not known (Norseth 1977).

D. Specific Organs and Systems

1. Lungs

Some authorities have described a characteristic pattern of bronchopneumonia, emphysema, and/or severe fibrosis in test animals from inhalation or intratracheal administration of powdered aluminum or aluminum compounds. Others believed inhaled aluminum compounds produced nonspecific, dust effects, which were reversible, and that inhaled aluminum compounds are useful in prophylaxis and treatment of silicosis. The gold mining and ceramic industries report the best results with aluminum prophylaxis. The disease entity "aluminosis" is not proved. However, there are numerous reports of lung fibrosis and pneumothorax in workers in aluminum smelting, in processes involving metallic aluminum dust, and in the manufacture of alumina abrasive. A high incidence of bronchial asthma was found in workers exposed to bentonite (an aluminosilicate clay). Aluminum powder inhalation in guinea pigs aggravated their pulmonary tuberculosis (Browning 1969). Asthma symptoms have also been reported in workers exposed to Al fumes in soldering and welding (Stokinger 1981).

2. Skin

Soluble Al salts cause an acid irritation from hydrolysis. Long contact with alum may produce a congestive, anesthetic condition in the fingers (acroanesthesia) (Stokinger 1981).

E. Teratogenicity

$AlCl_3$ (15 mg injected in the yolk sac on the 4th day) did not cause defects in the chick embryos (Ridgway and Karnofsky 1952; cited by Shepard 1980). No adverse fetal effects were observed in the offspring of rats given 0.1% $AlCl_3$ in their drinking water on days 6 to 19 of gestation (McCormack et al. 1978; cited by Shepard 1980).

F. Mutagenicity

No information.

G. Carcinogenicity

Elevated incidence of cancer in Al reduction plant workers (lymphomas, malignant tumors of the nervous system, leukemia) are ascribed to exposure to the polycyclic aromatic hydrocarbons in coal tar pitch volatiles (Stokinger 1981)

V. REFERENCES

ACGIH. 1983. TLVs[®] Threshold limit values for chemical substances and physical agents in the work environment with intended changes for 1983-84. Cincinnati, Ohio: American Conference of Governmental Industrial Hygienists.

Browning E. 1969. Toxicology of industrial metals.

Lauwerys RR. 1983. Chapter II. Biological monitoring of exposure to inorganic and organometallic substances. In: Industrial chemical exposure: Guidelines for biological monitoring. Davis, CA: Biomedical Publications, pp. 9-50.

Merck Index. 1983. 10th ed. Rahway, New Jersey: Merck & Co., Inc.

NAS. 1977. National Academy of Sciences. Drinking water and health. Vol. I. Washington, DC: Printing and Publishing Office National Academy of Sciences.

Norseth T. 1977. Aluminum. In: Toxicology of metals, Vol. II. Springfield VA: National Technical Information Service, pp. 4-14. PB-268-324.

Shepard TH. 1980. Catalog of teratogenic agents, 3rd ed. Baltimore: The Johns Hopkins University Press.

Stokinger HE. 1981. Chapter 29. The metals. In: Patty's industrial hygiene and toxicology. 3rd ed. Volume 2A, Toxicology. Clayton GD, Clayton FE, eds. New York: A Wiley-Interscience Publication. John Wiley and Sons, pp. 1493-2060.

Versieck J. 1984. Trace element analysis - A plea for accuracy. Trace Elements Med 1(1):2-12.

ANTIMONY

MAMMALIAN TOXICITY SUMMARY

I. INTRODUCTION

A. Occurrence and Production

Antimony is associated with the sulfide mineral stibnite and complex sulfide ores containing lead, copper, silver, and mercury. Resources in the United States are small. Antimony and antimony oxide recovery operations are or have been in Laredo, Texas; Kellogg, Idaho; Thompson Falls, Montana; El Paso, Texas; and Texas City, Texas (A. D. Little 1976).

B. Uses

Antimony is alloyed with lead, copper, and other metals. Certain compounds are used for fireproofing textiles and for ceramics and glassware, pigments, and antiparasitic drugs. Uses of alloys include ammunition, bearing metals, and lead storage batteries (USEPA 1980; Elinder and Friberg 1977).

C. Physical and Chemical Properties

Antimony (Sb), atomic no. 51, melts at 631°C and boils at 1750°C. It belongs to the same periodic group as arsenic and resembles it chemically and biologically. Antimony exhibits two valences in its compounds: +3 and +5 (Elinder and Friberg 1977). Most inorganic antimony compounds are water insoluble or decompose in water. Compounds with organic ligands, such as Sb K tartrate, are water soluble. Sb_2O_3 and Sb_2S_3 dissolve in strong alkali solution. Only Sb(III) exhibits definite cationic behavior (USEPA 1980).

II. EXPOSURE AND EXPOSURE LIMITS

A. Oral

Food is the primary exposure route for the normal population. Reported daily intakes range from 10 µg/day (determined by neutron activation analysis) to 250 to 1,250 µg/day (by atomic absorption spectrophotometry without extraction) (Elinder and Friberg 1977). The USEPA ambient water criterion for drinking water is 146 µg Sb/L (USEPA 1980).

B. Inhalation

Air concentrations in Chicago (reported in 1970) ranged from 1.4 to 55 ng/m^3 (average 32 ng/m^3). Tobacco contains 0.1 mg Sb/kg dry wt., about 20% of which is inhaled while smoking cigarettes (Elinder and Friberg 1977).

The threshold limit value as a time-weighted average in workroom air for an 8-h day is 0.5 mg/m^3 for antimony and compounds and for Sb_2O_3.* The TLV for stibine is also 0.5 mg/m^3. Its short-term exposure limit is 1.5 mg/m^3 (ACGIH 1983). The OSHA permissible exposure limit for antimony and compounds is 0.5 mg Sb/m^3 (OSHA 1981). Stibine (SbH_3) gas exposure may occur when lead-acid batteries are over charged (ACGIH 1980).

C. Dermal

Skin contact may occur with textiles flame-proofed by antimony compounds.

D. Total Body Burden and Balance Information

According to Snyder et al. (1975), the antimony balance for 70-kg reference man (µg/day) includes intake from food and fluids, $\sim$ 50, and airborne intake, 0.05 and losses in urine, $\sim$ 40, feces, $\sim$ 9, and hair, 1. The total Sb body burden in an average Japanese was $\sim$ 1 mg (Rhodamine B determination).

III. TOXICOKINETICS

A. Absorption

About 15% of a single oral dose of Sb K tartrate was found in the urine and tissues of mice; the amount of gastrointestinal excretion was not estimated. Hamsters receiving 0.43 mg Sb K tartrate i.p. excreted 50% of the dose in the feces within 24 h. No pertinent data were found for absorption from the lungs (Elinder and Friberg 1977).

B. Distribution

1. General

Highest tissue concentrations after acute or chronic exposure to antimony are usually in the thyroid, adrenals, liver, and kidney. In humans given intravenous injections of radioantimony (as Na Sb dimercaptosuccinate), liver, thyroid, and heart had the highest concentrations (Elinder and Friberg 1977).

2. Blood

One year after treatment for bilharzia, three subjects had an average 6.7 µg Sb/L in their blood (and 27.6 µg/L in their urine) compared to 3.4 and 6.2 µg/L in blood and urine, respectively, of three untreated subjects. Normals contain mean values of 3 µg Sb/L whole blood and 0.8 µg

---

* No exposure is recommended in Sb_2O_3 production based on the carcinogenic potential for humans.

Sb/L serum (Elinder and Friberg 1977). The range in serum or plasma of normals is 0.5 to 5 µg/L (Lauwerys 1983). Antimony levels in the blood of workers exposed to dusts of Sb, Sb_2O_3, Sb_2S_5 were 5 to 200 mg/L (urine values: 5 to 182 mg/L), but they do not seem to correlate well with the dustiness of operations (NIOSH 1978).

C. Excretion

Urinary excretion is apparently higher for As(V) than for As(III) compounds. Conversely, gastrointestinal excretion is higher for As(III) compounds. A small fraction of absorbed antimony has a long biological half-life in humans (Elinder and Friberg 1977).

Workers exposed to ~ 3 mg Sb/m^3 have 0.8 to 9.6 mg Sb/L in their urine, much higher than in the urine of normals (~ 0.5 to 2.6 µg/L) (Elinder and Friberg 1977). A biological threshold limit for antimony in urine has not been proposed due to lack of information (Lauwerys 1983).

IV. EFFECTS

The physiological effects of antimony have been reviewed by A. D. Little (1976), Elinder and Friberg (1977), USEPA (1980), and Stokinger (1981).

A. Acute and Other Short-Term Exposures

1. Inhalation

Acute effects in humans from inhalation exposure to $SbCl_3$ at 73 mg/m^3 were irritation and soreness of the upper respiratory tract. After exposure to unknown high concentrations of $SbCl_5$, three workers developed severe pulmonary edema (two died). Workers with heavy exposure to antimony fumes developed abdominal cramps, diarrhea, and vomiting. Stibine exposure causes headache, nausea, weakness, slow breathing, and weak pulse (Elinder and Friberg 1977).

Guinea pigs inhaling Sb_2O_3 at 45 mg/m^3 for 33 to 609 h showed signs of interstitial pneumonia. Rats and rabbits inhaling 3.1 and 5.6 mg Sb/m^3 as Sb_2S_3 for 6 wk developed parenchymatous degeneration of the myocardium (Elinder and Friberg 1977).

2. Oral

Children ingesting ~ 30 mg Sb/L in a contaminated lemon drink suffered vomiting, nausea, and diarrhea (Elinder and Friberg 1977).

3. Other routes

Intravenous injections of antimony compounds in laboratory animals cause an acute circulatory response and pathological changes in the electrocardiagram. Nausea and vomiting commonly occur during clinical treatment with antimony compounds (Elinder and Friberg 1977).

B. Chronic Exposure

Pneumoconiosis and sometimes obstructive lung diseases have been observed after long-term occupational exposure. Antimony trioxide exposure in the workplace has contributed to heart diseases (with fatal outcomes in some cases) as has the use of antimony compounds in the treatment of parasitic diseases. Liver damage and gastrointestinal ulcers are other signs of long-term exposures (Elinder and Friberg 1977). See part IV.D. for details.

Long-term exposure of rats fo 5 ppm Sb as Sb K tartrate in their drinking water shortened their lifespan by ~ 150% compared with that of the controls (Elinder and Friberg 1977).

C. Biochemistry

1. Effects on enzymes

Antimony compounds inhibit phosphofructokinase and, presumably, NADH-oxidase in vitro. In vivo, antimony compounds inhibit monoamine oxidase activity in brain and liver but increase cholinesterase activity in myocardium (USEPA 1980). Other antimony compounds inhibit succinoxidase and pyruvate brain oxidase. Sb interferes with cellular metabolism by combining with the sulfhydryl group of respiratory enzymes (Stokinger 1981).

2. Metabolism

No information.

3. Antagonisms and synergisms

No information.

D. Specific Organs and Systems

1. Respiratory system

Acute respiratory effects are described in Part IV.A. Antimony exposure among smelter workers (4.7 to 11.8 mg/m^3) caused respiratory symptoms: soreness and nosebleeds (> 70%), rhinitis (20%), pharyngitis (8%), pneumonitis (5.5%), and tracheitis (1%). Pneumoconiosis-like conditions appeared in chest x-rays. Among antimony trioxide workers, the incidences of pneumoconiosis and emphysema were 21% and 42%, respectively. Rats exposed to unstated concentrations of Sb_2O_3 for $\leq$ 14 mo developed pneumonitis, lipoid pneumonia, fibrous thickening of alveolar walls, and focal fibrosis (Elinder and Friberg 1977).

2. Cardiovascular system

Acute exposures of animals elicit a fall in blood pressure and pathological EKG changes, notably an inversion of the T-wave. This effect has been observed in long-term exposure to antimony medicinals. Tarter emetic may be involved in schistosomal myopathy; myopathy has been observed in rats and rabbits exposured to antimony (Elinder and Friberg 1977).

Occupational exposures to airborne Sb_2S_3 (> 3.0 mg/m^3) were associated with sudden deaths and heart complications.

In a Russian study, more than half of 8 antimony production workers complained of cardiac pain. There was diffuse myocardial damage and diminished contractile ability (USEPA 1980).

3. Liver and kidney

Acute exposure of rabbits to 60 mg Sb K tartrate caused fatty degeneration in the liver and in the convoluted tubules of the kidney. Liver damage, as shown by rises in serum GOT and GPT, have been observed in humans at the start of antimony therapy (Elinder and Friberg 1977).

4. Gastrointestinal system

Nausea and vomiting are common effects of clinical treatment with antimony. Six percent of 111 antimony workers had ulcers compared to 1.5% among 3,912 other employees (Elinder and Friberg 1977).

5. Reproductive system

A 1967 Russian study reported a 12% incidence of spontaneous late abortions among female antimony smelter workers compared to 4.1% in the controls (no effects are mentioned in the secondary reference). Birth weight was not different, but weight at 1 yr was significantly less. A 1955 study reported that premature deliveries occurred in 3.4% of women antimony workers compared to 1.2% in controls. Both groups had elevated incidences of gynecological disorders. Uterine and ovarian disorders have been observed in experimental animals (USEPA 1980).

6. Skin and eye

Antimony processing workers developed skin rashes on the forearms and thighs resembling chicken pox pustules. Antimony smelter workers exhibited vesicular dermal lesions with necrotic centers, which left hyperpigmented scars (could not be definitively ascribed to antimony because of other exposures). Sb_2O_3 is not a primary skin irritant or skin sensitizer. No irritation or systemic toxicity has been observed after application of Sb_2O_3 to the eyes or intact skin of animals (USEPA 1980).

E. Teratogenicity

No fetal malformations were observed after pregnant rats received 125 or 250 mg Sb/kg (route?) as Sb dextran glycoside on 5 days during days 8 to 14 of gestation (USEPA 1980).

F. Mutagenicity

Sb_2O_3, $SbCl_3$, and $SbCl_5$ were mutagenic in Bacillus subtilis. Tartar emetic at $\geq 10^{-8}$ M significantly reduced the mitotic index of cultured human leukocytes and increased the number of chromatid breaks in chromosomes (USEPA 1980).

G. Carcinogenicity

Between 1925 and 1971, 15 deaths occurred in a total work force of 1,081 among workers at an antimony and zircon works in the United Kingdom who had been exposed for 7 to 43 yr (average 22 yr). Only two were nonsmokers and one, a "very light smoker." The rate was about twice ad high as the local death rate for lung cancer (Stokinger 1981).

V. REFERENCES

ACGIH. 1980. Documentation of the threshold limit values. 4th ed. Cincinnati, Ohio: American Conference of Governmental Industrial Hygienists, Inc.

ACGIH. 1983. TLVs® Threshold limit values for chemical substances and physical agents in the work environment with intended changes for 1983-84. Cincinnati, Ohio: American Conference of Governmental Industrial Hygienists.

AD Little. 1976. Literature study of selected potential environmental contaminants: Antimony and its compounds. Washington, D.C.: Office of Toxic Substances, U.S. Environmental Protection Agency. EPA-560/2-76-002.

Elinder C-G, Friberg L. 1977. Antimony. In: Toxicology of metals--Volume II. Springfield, Virginia. National Technical Information Service, pp. 15-29. PB-268 324.

Lauwerys RR. 1983. Chapter II. Biological monitoring of exposure to inorganic and organometallic substances. In: Industrial chemical exposure: Guidelines for biological monitoring. Davis, CA: Biomedical Publications, pp. 9-50.

NIOSH. 1978. Natl. Occup Saf. Health Admin. Criteria for a recommended standard... Occupational exposure to antimony. Washington, DC: US Govt. Printing Office.

OSHA. 1981. Occupational Safety and Health Standards. Subpart 2--Toxic and hazardous substances. Code of federal regulations 29 (Part 1910.1000): 673-679.

Stokinger HE. 1981. Chapter 29, The metals. In: Patty's industrial hygiene and toxicology, 3rd ed., Volume IIA, Toxicology. Clayton DG, Clayton FE, eds. New York: Wiley-Interscience, John Wiley & Sons, pp. 1493-2060.

USEPA. 1980. Environmental Criteria and Assessment Office, U.S. Environmental Protection Agency, Ambient water quality criteria for antimony, Springfield, VA: National Technical Information Service, PB81-117319.

ARSENIC

MAMMALIAN TOXICITY SUMMARY

I.   INTRODUCTION

A. Occurrence and Production

Arsenic is widely distributed in the earth's crust, with an abundance of about 5 ppm. It is found in minerals such as arsenopyrite ($FeAs_2 \cdot FeS_2$), enargite ($3Cu_2S \cdot As_2S_5$), and orpiment (As_2S_3), most often in the trivalent form. It is produced as a by-product from gold, copper, and lead smelters. Roasting the ores produces gaseous arsenic, which is precipitated with the dust as As_2O_3. The trioxide is then purified and becomes the usual article of commerce (Stokinger 1981; Tsuchiya et al. 1977).

B. Uses

Arsenic compounds have been used therapeutically for over two millenia. The better known uses include treatment of dermatoses and syphilis. A vestige of these uses continues in agriculture (the metal in sheep dip, sodium arsenite as a weed killer), silviculture (cacodylic acid and its sodium salt as silvicides) and the preparation of their products (arsenic trioxide in tanning, taxidermy and wood preservation). Small quantities of organic arsenic compounds such as arsanilic acid are used as a growth promoter (essential trace element) for swine and poultry.

Arsenic is alloyed to lead to harden it for many purposes, including battery grids, cable sheathing, solder, shot, and bearings. It is used in glass manufacture and in semiconductors (Stokinger 1981; Tsuchiya et al. 1977).

C. Physical and Chemical Properties

Strictly speaking, arsenic is a metalloid, not a true metal, as shown by the formation of compounds like arsine (AsH_3), a gas, and $AsCl_3$, a liquid. Inorganic compounds have valences of +3 (e.g., the trioxide, As_2O_3) or +5 (the pentoxide, As_2O_5). These oxides form acids and salts (arsenites and arsenates). Most inorganic arsenic compounds, except heavy metal salts, are reasonably soluble (Stokinger 1981).

II. EXPOSURE AND EXPOSURE LIMITS

A. Oral

As would be expected from its ubiquitous distribution, traces of arsenic are usually found in water and food. Generally speaking, the highest concentrations are in seafood (Tsuchiya et al. 1977). Estimates of dietary arsenic range from 0.15 to 0.40 mg/person/day.

Arsenic was found in the finished water of 83 U.S. cities at levels from < 1.0 to 50 µg/L (USEPA 1975). However, some smaller systems contain more (up to 393 µg/L in the communities selected by Valentine et al. [1979]). The national interim primary drinking water standard is 50 µg/L (USEPA 1975). A more recent ambient water quality criteria document (USEPA 1980), recommends 2.2 ng/L, to give a negligible (10^{-6}) incremented increase of cancer risk.

B. Inhalation

Arsenic is released to air by the smelting of ores containing it, by combustion of coal, and by the use of arsenical pesticides. The average level in the U.S. is 0.2 µg/m³, but has been as high as 2.5 µg/m³ near a copper smelter (Tsuchiya et al. 1977). Arsenic levels in the urine and handwashings of schoolchildren and in playground dirt and dust correlated inversely with distance from a copper-lead smelter (Buchet et al. 1980).

The threshold limit value as a time-weighted average in workroom air for an 8-h day is 0.2 mg As/m³ for arsenic metal and soluble compounds (ACGIH 1983). The OSHA permissible exposure limit is 0.5 mg As/m³ for organic arsenic and 0.01 mg/m³ for inorganic arsenic (OSHA 1981), but NIOSH (1975) has recommended 0.002 mg/m³ as the limit.

C. Dermal

Although dermatotoxicity is well known, there is no evidence for direct dermal absorption of arsenic.

D. Total Body Burden and Balance Information

According to Snyder et al. (1975), the arsenic balance for the 70-kg reference man (in mg/day) is: intake from food and fluids 1.0 and airborne intake 1.4×10^{-3}; losses from urine 0.05, feces 0.8, and others < 0.16.

III. TOXICOKINETICS

A. Absorption

Absorption can be highly dependent on the chemical form and (for inhalation) the physical form of the arsenic. However, the common forms (particularly arsenic trioxide, the most prevalent) are well absorbed by inhalation and (essentially completely) from the gastrointestinal tract (Tsuchiya et al. 1977; USEPA 1980).

B. Distribution

1. General

Once absorbed, the arsenic is widely distributed to the tissues by the blood. The details may be obscured by the biotransformation of the arsenic, particularly between valence states. Arsenic is quickly taken up (except in the rat) by many tissues, but particularly by the liver, kidney, lungs, spleen, and skin. Bone and muscle have lower concentrations, but they and skin become the major depots of the body burden (Tsuchiya et al. 1977; USEPA 1980).

2. Blood

Blood levels do not correlate well with exposure levels; hair and urine (excretion routes) are much more useful as biological monitors of exposure. Mean blood levels were 0.29 to 0.51 µg As/L for populations exposed to drinking water containing 6 to 123 µg As/L, but the concentrations were not dose related. The mean blood level was 1.33 ± 1.18 µg/L for a population drinking 393 µg As/L (Valentine et al. 1979).

3. Adipose

Arsenic may accumulate in fat including neural tissue components (e.g., myelin) that are high in lipids, phospholipids, or phosphatides (USEPA 1980).

C. Excretion

Arsenic is excreted in the urine, the feces, and by the dermis as shed skin, hairs, and nails. The nails and hair may be used to detect past, discontinued exposure (Tsuchiya et al. 1977; USEPA 1980). The method of choice for monitoring industrial exposure to inorganic arsenic is determination of inorganic arsenic and its metabolites monomethylarsonic acid and cacodylic acid (Lauwerys 1983). Normal arsenic concentrations in urine of occupationally unexposed adults are usually < 100 µg/L (Stokinger 1981).

IV. EFFECTS

A. Acute and Other Short-Term Exposures

Acute effects are generally seen only after a large overdose, whether accidental, suicidal, or homicidal. The main effects are seen in the gastrointestinal system. These include throat constriction followed by difficulty in swallowing, abdominal discomfort, nausea, chest pain, vomiting, and watery diarrhea. Larger doses produce systemic collapse due to severe hypotension, restlessness, convulsions, coma and death, often by cardiac failure (USEPA 1980; Tsuchiya et al. 1977).

A few compounds have their own peculiar effects. Arsine (AsH_3) causes extensive hemolysis, which leads to jaundice, pigmented urine, then anuria, severe anemia, and death from myocardial failure due to anemia. Some common compounds (arsenic trioxide, arsenic trichloride) and the obsolete arsenical war gases (such as Lewisite, dichloro[2-chlorovinyl]arsine) are irritants to the lungs and exposed skin (Stokinger 1981; Tsuchiya et al. 1977).

B. Chronic Exposure

In most cases, effects (including carcinogenesis, cardiovascular disease, and neurological effects) are seen only after chronic low-dose exposures, whether environmental or occupational. Specific data on threshold effects are limited, due to poor documentation of the actual exposure. [These disorders have been linked to exposure to drinking water containing > 50 μg As/L (Valentine et al. 1979).] Many of the controlled studies are in the rat, which happens to be a poor model due to its aberrant distribution patterns (USEPA 1980). Details of these affected systems are discussed below.

The skin is the major organ of interest, since most effects are seen here--even after oral ingestion of the arsenic. Effects on the mucous membrane, peripheral nervous system, and gastrointestinal system are quite common. Rarer are effects on the blood, heart, and liver (Stokinger 1981; Tsuchiya et al. 1977).

C. Biochemistry

1. Effects on Enzymes

It is generally accepted that most, if not all, of the effects of arsenic are mediated via the reaction between the metal and thiol groups of various enzymes and cofactors (such as glutathione). The details are quite complex [USEPA (1980) and Stokinger (1981) are leading references]. This complexity has been increased by our knowledge of interactions with selenium, particularly in carcinogenesis (Stokinger 1981).

2. Metabolism

The major metabolism reaction is methylation to dimethylarsinic acid [cacodylic acid] and monomethylarsonic acid, thus detoxifying the arsenic. This is a first-order reaction, but can be saturated by large doses (USEPA 1980).

3. Antagonisms and Synergisms

The most significant antagonism is the use of BAL ("British Anti-Lewisite," 2,3-dimercaptopropanol) as an antidote to arsenical poisoning. The highly reactive thiol groups form cyclic dithioarsenites, which are more stable than the protein-thiol arsenites, thus preventing the enzyme effects. Also, selenium is capable of reducing the carcinogenicity of arsenic, perhaps even eliminating it (Stokinger 1981). No synergisms are known.

4. Physiological Requirements

There is evidence that arsenic is a required trace element for some domestic animals. It is a feed additive for swine and poultry (Tsuchiya et al. 1977).

D. Specific Organs and Systems

    1. Skin

        Skin lesions are the classic effects of arsenic toxicity. Acute
dermatitis is usually seen first. This begins as local erythema with burning
and itching, giving the skin a mottled appearance. This is followed by
swelling and sometimes papular or vesicular eruptions. Some of these lesions
may be due to local irritation or to allergic contact dermatitis.

        Chronic dermatotoxicity is quite distinctive. Melanosis, due to
arsenic's modification of cell metabolism, is first seen on the eyelids,
temples, neck, areolae of the nipples and in the folds of the axillae. It
may spread throughout the trunk. It is often accompanied by hyperkeratosis,
hyperhidrosis (excessive sweating on palms and soles), and warts
(Stokinger 1981; Tsuchiya et al. 1977).

    2. Mucous Membranes

        The acute effects on the skin are usually accompanied by similar
irritation effects on the mucous membranes of the eyes (conjunctivitis), nose
(rhinitis), pharynx, and bronchial passages. Corneal anesthesia and
consequent corneal ulcer have been seen, as well as perforation of the nasal
septum (Stokinger 1981; Tsuchiya et al. 1977).

    3. Peripheral Nervous System

        Peripheral neuritis, primarily of the upper and lower extremities,
is commonly seen. It generally begins as sensory disturbances (numbness,
tingling, "pins and needles" sensation--parathesias, pain, burning
tenderness) and progresses to muscular weakness, even near paralysis.
Effects may be unilateral or bilateral and are associated with neuronal
degeneration, which is believed to be due to inhibition of some enzymes
(Stokinger 1981; Tsuchiya et al. 1977).

    4. Gastrointestinal System

        Chronic arsenic toxicity includes loss of appetite and
gastrointestinal disturbances. Frank gastroenteritis is rare, as is
cirrhosis and other liver toxicities (Stokinger 1981; Tsuchiya et al. 1977).

    5. Blood and Cardiovascular System

        Anemia and leukopenia are sometimes seen; this effect is due to
metabolic disturbances and is not related to the arsine-induced hemolysis
discussed above. There are some reports of myocardial injury in arsenic
poisoning (Stokinger 1981; Tsuchiya et al. 1977).

E. Teratogenicity

        Arsenic crosses the placenta and is a known animal teratogen
(USEPA 1980; Tsuchiya et al. 1977). Limited epidemiological data suggest
that similar effects occur in humans (IARC 1980).

F. Mutagenicity

In a number of animal and human studies, chronic arsenic toxicity is associated with chromosome aberrations. This seems to be due to interference with normal DNA repair processes, presumably by the usual biochemical mechanism (USEPA 1980; Tsuchiya et al. 1977).

G. Carcinogenicity

"There is sufficient evidence that inorganic arsenic compounds are skin and lung carcinogens in humans" (IARC 1980). The evidence is epidemiological studies of humans occupationally and environmentally exposed. Animal studies have been negative, perhaps due to selenium antagonism (Stokinger 1981).

V. REFERENCES

ACGIH. 1983. TLVs® Threshold limit values for chemical substances and physical agents in the work environment with intended changes for 1983-84. Cincinnati, Ohio: American Conference of Governmental Industrial Hygienists.

Buchet JP, Roels H, Lauwerys R, et al. 1980. Repeated surveillance of exposure to cadmium, manganese and arsenic in school-age children living in rural, urban and nonferrous smelter areas in Belgium. Environ Res 22:95-108.

IARC. 1980. International Agency for Research on Cancer. Some metals and metallic compounds, arsenic and arsenic compounds. IARC Monogr Eval Carcinog Risk Chem Humans 23:39-141.

Lauwerys RR. 1983. Chapter II. Biological monitoring of exposure to inorganic and organometallic substances. In: Industrial chemical exposure: Guidelines for biological monitoring. Davis, CA: Biomedical Publications, pp. 9-50.

NAS. 1980. National Academy of Sciences. Recommended dietary allowances. 9th ed. Washington, DC: Printing and Publishing Office, National Academy of Sciences.

NIOSH. 1975. Natl. Inst. Occupational Safety and Health. Criteria for a recommended standard: occupational exposure to inorganic arsenic. Washington, DC: U.S. Dept. Health, Education, and Welfare, Public Health Service, Center for Disease Control.

OSHA. 1981. Occupational Safety and Health Admin. Occupational safety and health standards. Subpart 2--Toxic and hazardous substances. Code of federal regulations 29 (Part 1910.1000) pp. 673-679.

Snyder WS, Cook MJ, Nasset ES, Karhausen LR, Howells GP, Tipton IH. 1975. International Commission on Radiological Protection. Report of the task group on reference man. New York. ICRP Publication 23.

Stokinger HE. 1981. Chapter 29. The metals. In: Patty's industrial hygiene and toxicology. 3rd ed. Volume 2A. Clayton GD, Clayton FE, eds. New York: A Wiley-Interscience Publication. John Wiley and Sons, pp. 1493-2060.

Tsuchiya K, Ishinishi N, Fowler BA. 1977. In: Toxicology of metals. Volume II. Springfield, VA: National Technical Information Service, pp. 30-70. PB-268-324.

USEPA. 1975. U.S. Environmental Protection Agency. National interim primary drinking water regulations. (40 FR 59566).

USEPA. 1980. U.S. Environmental Protection Agency. Ambient water quality criteria for arsenic. Springfield, VA: National Technical Information Service, No. PB81-117327.

Valentine JL, Kang HK, Spivey, G. 1979. Arsenic levels in human blood, urine and hair in response to exposure via drinking water. Environ Res 20:24-32.

BARIUM

MAMMALIAN TOXICITY SUMMARY

I. INTRODUCTION

   A. Occurrence and Production

        Barium is produced by mining barite (barium sulfate [$BaSO_4$]); the ore used is 75 to 98% barium. Ore-grade barite is found in many parts of North America, but Nevada supplies 50% of United States production (Stokinger 1981).

   B. Uses

        Over 90% of barium use is in the form of ground barite used in oil and gas well drilling. The rest is used to produce lithopone (ZnS + $BaSO_4$) and barium chemicals. The latter are used in the glass, paint, and rubber industries. Barium is also used to remove gas from vacuum tubes. $BaCO_3$ has been used as a rat poison. Other uses for barium compounds include refractories, flares, and fireworks (Reeves 1977; Stokinger 1981; Browning 1969).

   C. Chemical and Physical Properties

        Barium, atomic no. 56, melts at 752°C and boils at 1640°C. It is soluble in alcohol and insoluble in benzene. In its chemistry, barium is a typical alkaline earth element, closely resembling calcium (Stokinger 1981).

II. EXPOSURE AND EXPOSURE LIMITS

   A. Oral

        Small proportions of barium accompany calcium in virtually every food stuff. It is estimated that the average intake is 1.33 mg/day (Reeves 1977). Barium was found in the finished water of the 100 largest U.S. cities at a median concentration of 43 µg/L; the maximum was 380 µg/L and 94% were less than 100 µg/L (Durfor 1964; cited by NAS 1977). Another study found barium in 99.7% of the finished-water systems sampled, with a mean level of 28.6 µg/L and a range of 1 to 172 µg/L (Kopp 1970; cited by NAS 1977). The maximum level found in 2,595 tap water samples was 1,550 µg/L (McCabe 1970; cited by NAS 1977).

        The national interim primary drinking water standard for barium is 1 mg/L (USEPA 1975).

   B. Inhalation

        The threshold limit value as a time-weighted average in workroom air for an 8-h day is 0.5 g Ba/m³ for soluble barium compounds (ACGIH 1983). The OSHA permissible exposure limit is 0.5 mg/m³ (OSHA 1981).

34

Barium may be a common constituent of urban air. Its compounds have been used as diesel fuel smoke suppressants and as automotive lubricants (Smith et al. 1975).

C. Dermal

Soluble salts are skin and mucous membrane irritants. The Ba dispersant in automotive lubricants is a mild eye irritant when used full strength (Smith et al. 1975).

D. Total Body Burden and Balance Information

According to Snyder et al. (1975), the barium balance for the 70-kg reference man in µg/day is: intake 750 from food and fluids and 0.09 to 26 x 10^{-3} from air; losses 0.05 in urine, 0.69 in feces, and 0.085 by other routes. They estimate the body burden is 22 mg (20 mg in skeleton). It has been estimated that an average skeleton contains 2 mg/kg of barium.

III. TOXICOKINETICS

A. Absorption

Soluble barium compounds are rapidly absorbed from the lungs and gastrointestinal tract. Insoluble compounds are not; hence barium sulfate is used as an x-ray contrast medium for gastrointestinal examination, despite its abominable chalky taste (Reeves 1977).

B. Distribution

1. General

Once absorbed, barium is accumulated in the bones. One study reported an intermediate depot in the muscles (Reeves 1977; Stokinger 1981). Lifetime accumulations are 7 ppm (ash) in bone and 11 to 24 ppm (ash) in soft tissues (Stokinger 1981).

2. Blood

Barium disappears from the blood within 24 h (Stokinger 1981). Normal blood has 0.08 to 0.4 mg/L, mostly in the plasma (Reeves 1977).

C. Excretion

Excretion is in the feces and (secondarily) urine, sweat, and other routes (Reeves 1977).

IV. EFFECTS

A. Acute and Other Short-Term Exposures

Soluble barium compounds (nitrate, sulfide, carbonate, chloride) have been involved in accidental and suicidal poisonings. Initial symptoms are gastrointestinal: nausea, vomiting, colic, and diarrhea.   Next comes myocardial and general muscular stimulation with tingling in the extremities. Severe cases continue to loss of tendon reflexes, heart fibrillation, general muscular paralysis, and death from respiratory arrest (Reeves 1977).

A fatal dose of $BaCl_2$ for a human is reported to be 0.8 to 0.9 g (0.55 to 0.6 g Ba) (Sollman 1953; cited by Stokinger 1981).   Symptoms include excessive salivation, vomiting, colic, diarrhea, convulsive tremors, slow hard pulse, elevated blood pressure, internal hemorrhages, and muscular paralysis.   Death may occur within a few hours or days (Stokinger 1981).

B.   Chronic Exposure

A benign pneumoconiosis, known as baritosis, has been reported by several studies as occurring in workers exposed to finely ground $BaSO_4$.   It consists of numerous evenly distributed nodules in the lungs that cause no specific symptoms or changes in pulmonary function.   Nodules usually disappear after cessation of exposure, but bronchial irritation may persist (Stokinger 1981).

In guinea pigs, subcutaneous exposure to $BaCl_2$ at 12 mg/kg daily for 26 wk affected the spleen, liver, and bone marrow, causing cellular changes in the blood.   Bone marrow lesions appeared in half the animals (Truhaut et al. 1958; cited by Stokinger 1981).

In rabbits injected repeatedly [intervals not stated] with 2 to 10 mg/kg for 98 to 193 days, central nervous system effects were reported (Stokinger 1981).

C.   Biochemistry

1.   Effects on Enzymes

Barium can activate the secretion of catecholamines from the adrenal medulla (Douglass and Rubin 1964; cited by Stokinger 1981).

2.   Metabolism

Barium is not metabolized.

3.   Antagonisms and Synergisms

Barium is mutually antagonistic to all muscular depressants (Zolessi and Stazzi 1939; cited by Stokinger 1981).   Barium is a physiological antagonist of potassium.   It has been confirmed that potassium deficiency occurs with barium poisoning (Lydlin et al. 1965; cited by Stokinger 1981).

D.   Specific Organs and Systems

1.   Muscle

The toxic effects of ingested barium are due to its muscle effects--stimulation followed by paralysis.   The probable mechanism is by transferring potassium to intracellular compartments, thus inducing hypokalemia with characteristic electrocardiogram effects (Reeves 1977).

2. Other

Effects in the hematopoietic systems and resulting changes in blood have been reported in guinea pigs, and changes have been reported in lungs of workers exposed to barium dust. See Section III, Chronic Exposure.

E. Teratogenicity

Injecting 20 mg $BaCl_2$ in to the chick yolk sac on the 8th day of incubation produced curled toes in about half of the survivors. Earlier injections did not have a teratogenic effect (Ridgway and Karnofsky 1952; cited by Shepard 1980).

F. Mutagenicity

No information.

G. Carcinogencity

Rats injected intratracheally with radioactive particles of $BaSO_4$ (^{35}S) developed bronchogenic carcinoma (squamous-cell type) (Amber and Watson 1958; cited by Stokinger 1981).

V. REFERENCES

ACGIH. 1983. TLVs[®] Threshold limit values for chemical substances and physical agents in the work environment with intended changes for 1983-84. Cincinnati, Ohio: American Conference of Governmental Industrial Hygienists.

Browning E. 1969. Toxicity of Industrial Metals. 2nd ed. London: Butterworths.

NAS. 1977. National Academy of Sciences. Drinking water and health. Vol. I. Washington, DC.

OSHA. 1981. Occupational Safety and Health Admin. Occupational safety and health standards. Subpart 2--Toxic and hazardous substances. Code of federal regulations 29 (Part 1910.1000) pp. 673-679.

Reeves AL. 1977. Barium. In: Toxicology of metals - Vol. II. Springfield, VA: National Technical Information Service, pp. 71-84, PB-268 324.

Shepard TH. 1980. Catalog of teratogenic agents, 3rd ed. Baltimore: The Johns Hopkins University Press.

Smith, IC, Ferguson TL, Carson BL. 1975. Metals in new and used petroleum products: Quantities and consequences. In: The role of trace metals in petroleum. Yen, TF, Ed. Ann Arbor, MI: Ann Arbor Science Publishers, pp. 123-148.

Snyder WS, Cook MJ, Nasset ES, Karhausen LR, Howells GP, Tipton IH. 1975. International Commission on Radiological Protection. Report of the task group on reference man. New York. ICRP Publication 23.

Stokinger HE. 1981. Chapter 29. The metals. In: Patty's industrial hygiene and toxicology. 3rd ed. Volume 2A. Clayton GD, Clayton FE, eds. New York: A Wiley-Interscience Publication. John Wiley and Sons, pp. 1493-2060.

USEPA. 1975. U.S. Environmental Protection Agency. National interim primary drinking water regulations. (40 FR 59566).

BERYLLIUM

MAMMALIAN TOXICITY SUMMARY

I. INTRODUCTION

A. Occurrence and Production

Beryllium is found in a few minerals. The commercial ores are beryl ($Be_3[AlSi_3O_9]_2$) and bertrandite ($Be_4[H_2Si_2O_9]$). It occurs at about 6 ppm in the earth's crust. By various processes the ores are converted to the oxide (BeO) or fluoride (BeF_2), which are reduced to the metal (Stokinger 1981; IARC 1980; Reeves 1977).

B. Uses

Beryllium is a "high-tech" metal, used only where its peculiar properties justify the high cost. Zinc beryllium silicate is used as a phosphor in fluorescent lights and similar products. Beryllium oxide is used in ceramics, heatsinks, as well as insulators for semiconductor applications, microwave tubes, etc., and as a moderator and/or reflector in nuclear reactors. Beryllium metal and alloys with aluminum and copper are used in electrical uses, reactors, aerospace components, etc. (IARC 1980; Stokinger 1981).

C. Chemical and Physical Properties

Beryllium is the only stable, lightweight metal with a high melting point; most of its uses are a consequence of this. At elevated temperatures, it is oxidized to the +2 oxidation state; subsequent reactions are very similar to those of trivalent aluminum (Stokinger 1981).

II. EXPOSURE AND EXPOSURE LIMITS

A. Oral

There is a small normal intake of beryllium from water and soil (via food), estimated at 100 µg/day (Reeves 1977). It has been found in U.S. drinking waters at 0.01 to 0.7 µg/L, with a mean of 0.013 µg/L (APHA 1976; cited by NAS 1977). USEPA (1980) gave criteria of 37, 3.7, and 0.37 ng Be/L drinking water which may result in incremental increases of cancer risk over a lifetime of 1 in 100,000; 1 in one million; and 1 in 10 million, respectively

B. Inhalation

There is some airborne beryllium, due to coal combustion, cigarette smoke and, in a few areas, beryllium processing plants (Reeves 1977; IARC 1980). The threshold limit value as time-weighted average in workroom air for an 8-h day is 0.002 mg/m³ for beryllium and its compounds

(ACGIH 1983). The OSHA permissible exposure limit is 0.002 mg/m³, with a peak limit of 0.025 mg/m³ over 30 min (OSHA 1981).

C. Dermal

There is no evidence for significant dermal absorption (Reeves 1977).

D. Total Body Burden and Balance Information

According to Snyder et al. (1975), the beryllium balance for 70-kg reference man in μg/day is intake from food and fluids 12 and from air < 0.01: losses from urine 1.0, from feces 10, and by other routes 1.0. The estimated body burden was 36 μg (27 μg in soft tissues).

III. TOXICOKINETICS

A. Absorption

Oral absorption estimates range from 0.001 to 0.01% (Snyder et al. 1975). Even highly soluble compounds are not well absorbed. Absorption from the lungs is slow (Stokinger 1981).

B. Distribution

1. General

After absorption, beryllium does not localize in the lungs but is widely distributed among the organs. Significant concentrations occur in the liver and the skeleton as part of the mineral substance (Reeves 1977; IARC 1980).

2. Blood

Snyder et al. (1975) estimated < 0.52 μg in the whole blood of 70-kg reference man. Typical blood concentrations of exposed workers were not found in the secondary references consulted. Transport in the blood and lymph is believed to be chiefly as an orthophosphate colloid (Stokinger 1981). Qualitative blood and urine tests can be done to confirm beryllium exposure, but concentrations have not been correlated with amounts due to recent exposure and to body stores. Published data do not adequately characterize the degree of exposure for biological values reported (Lauwerys 1983).

C. Excretion

Beryllium is excreted in the urine (ionic forms) and feces (colloidal forms collect in the liver). Elimination is extremely slow (IARC 1980; Stokinger 1981). Beryllium in the urine may reflect recent exposure, but increased beryllium in the urine can be detected for years after removal from exposure (Lauwerys 1983).

IV. EFFECTS

A. Acute and Other Short-Term Exposures

Only the soluble beryllium salts (such as the sulfate and fluoride) have acute effects (Stokinger 1981), which are primarily at the site of contact.

On the skin, one sees contact dermatitits--edematous papulovesicular lesions, which may be allergic. This is often accompanied by conjunctivitis. If beryllium compounds are embedded in the skin (such as after an injury from a broken fluorescent tube), slowly healing necrotizing granulomatous ulcerations develop.

Inhaled beryllium produces irritation of the entire respiratory tract: rhinitis, pharyngitis, tracheobronchitis, and, finally, a severe chemical pneumonitis. In severe cases, the latter develops into a fatal pulmonary edema (NIOSH 1972; Reeves 1977).

B. Chronic Exposure

1. Berylliosis

Chronic exposure to beryllium and its compounds can produce a frequently fatal pulmonary granulomatosis called "berylliosis." Great variation in individual susceptibility occurs. The major signs and symptoms include pneumonitis with accompanying cough, chest pain and general weakness, and, often, pulmonary dysfunction. Onset is insidious, with shortness of breath the first symptom, and miliary mottling of the x-rays and interstitial granulomatosis of pulmonary biopsies confirming the diagnosis. Accompanying effects include right heart enlargement leading to congestive heart failure, enlargement of the liver and spleen, cyanosis, finger "clubbing," kidney stones, and assorted biochemical abnormalities (NIOSH 1972; Reeves 1977; Stokinger 1981).

2. Other Chronic Effects

Animal studies have shown a variety of lesions not reported in man. These include pulmonary cancer, osteosarcoma, rickets (apparently due to intestinal precipitation and nonabsorption of beryllium phosphate), macrocytic anemia, and alveolar metaplasia (NIOSH 1972; Reeves 1977).

C. Biochemistry

1. Effects on Enzymes

Beryllium inhibits a number of enzymes, especially in vitro. The toxicological significance of these effects is unknown (Reeves 1977; IARC 1980). The toxic effects of beryllium may be ascribed to damage of lysosomes, which release cell-destroying enzymes. Beryllium in vitro inhibits Mg-activated alkaline phosphatase, Ca-activated adenosine triphosphatase, and hyaluronidase. It activates succinic dehydrogenase (Stokinger 1981).

2. Metabolism

There is no evidence of metabolism of beryllium, other than (perhaps) limited oxidation of the metal to the divalent state, which is absorbed. Beryllium is involved internally in a number of ill-defined complexes (Reeves 1977). The principal form in which beryllium is transported may be an orthophosphate colloid (Stokinger 1981).

3. Antagonisms and Synergisms

None known.

D. Specific Organs and Systems

1. Immune System

Berylliosis seems to involve an interference with the immune system. The exact mechanism is obscure (Reeves 1977; Stokinger 1981). Beryllium is quite antigenic, acting as a hapten to provoke an antigenic response. This then leads to inflammatory responses, including many of the specific components of berylliosis. A recent study (Stiefel et al. 1980) noted a beryllium-induced increase of T-lymphocytes, but the significance of this is unknown.

The toxic mechanism, other than this immune effect, and direct irritation (acute effects), is unknown.

2. Skin

Contact dermatitis and granulamatous ulcerations from embedded particles may occur.

E. Teratogenicity

No positive studies were found. Up to 113 μmol $BeSO_4$ produced no defects in the chick embryo (Ridgway and Karnofsky 1952; cited by Shepard 1980).

F. Mutagenicity

Beryllium compounds showed mixed results in various in vitro mutagenicity tests (IARC 1980).

G. Carcinogenicity

Beryllium is carcinogenic in three animal species (rats, mice, rabbits), but human evidence (occupational epidemiologic studies) is inadequate to confirm this (IARC 1980). One peculiarity of the most extensive study is that an excess of lung tumors occurs in those ex-workers with least exposure (< 5 yr work) but not in those who worked longer (Stokinger 1981).

V. REFERENCES

ACGIH. 1983. TLVs® Threshold limit values for chemical substances and physical agents in the work environment with intended changes for 1983-84. Cincinnati, Ohio: American Conference of Governmental Industrial Hygienists.

IARC. 1980. Internat. Agency for Research on Cancer. Monographs on the evaluation of the carcinogenic risk of chemicals to humans. Beryllium and beryllium compounds. IARC Monogr Eval Carcinog Risk Chem Humans 23:143-204.

Lauwerys RR. 1983. Chapter II. Biological monitoring of exposure to inorganic and organometallic substances. In: Industrial chemical exposure: Guidelines for biological monitoring. Davis, CA: Biomedical Publications, pp. 9-50.

NAS. 1977. National Academy of Sciences. Drinking water and health. Vol. I. Washington, DC.

NIOSH. 1972. Natl. Inst. Occupational Safety and Health. Criteria for a recommended standard: occupational exposure to beryllium. Washington, DC: National Institute for Occupational Safety and Health. PB HSM 72-10268.

OSHA. 1981. Occupational Safety and Health Admin. Occupational safety and health standards. Subpart 2--Toxic and hazardous substances. Code of federal regulations 29 (Part 1910.1000) pp. 673679.

Reeves AL. 1977. In: Toxicology of metals - Vol. II. Springfield, VA: National Technical Information Service, pp. 85-109, PB-268 324.

Shepard TH. 1980. Catalog of teratogenic agents, 3rd ed. Baltimore: The Johns Hopkins University Press.

Snyder WS, Cook MJ, Nasset ES, Karhausen LR, Howells GP, Tipton IH. 1975. International Commission on Radiological Protection. Report of the task group on reference man. New York. ICRP Publication 23.

Stiefel T, Schulze K, Zorn H, Toelg G. 1980. Toxicokinetic and toxicodynamic studies of beryllium. Arch Toxicol 45:81-92.

Stokinger HE. 1981. Chapter 29. The metals. In: Patty's industrial hygiene and toxicology. 3rd ed. Volume 2A. Clayton GD, Clayton FE, eds. New York: A Wiley-Interscience Publication. John Wiley and Sons, pp. 1493-2060.

USEPA. 1980. U.S. Environmental Protection Agency. Ambient water quality criteria for beryllium. Springfield, Virginia: National Technical Information Service. PB81-117350.

BISMUTH

MAMMALIAN TOXICITY SUMMARY

I. INTRODUCTION

A. Occurrence and Production

Bismuth is mainly found as the free metal and in the mineral bismuthinite (Bi_2S_3). The primary U.S. source is as a by-product of the refining of lead and copper ores (Stokinger 1981).

B. Uses

Bismuth is used to produce low-melting alloys for use in fusible elements of specialized products such as automatic sprinklers. It is also added to steel and iron to produce castings that can be machined more easily. In the past, it was used in numerous pharmaceuticals for the treatment of syphillis and skin and digestive tract ailments. The oxide and nitrate forms are widely used in glass and ceramics manufacture (Stokinger 1981).

C. Chemical and Physical Properties

Bismuth, atomic no. 83, melts at 271.3°C and boils at 1630°C. It is insoluble in water and soluble in hot H_2SO_4 and HNO_3. It expands on solidification and has the lowest thermal conductivity of all metals except mercury. Common compounds are water insoluble; BiF_5 reacts violently with water (Stokinger 1981).

II. EXPOSURE AND EXPOSURE LIMITS

A. Oral

Some United States drinking waters contain an average 0.01 mg Bi/L. No available information was found for bismuth in food (Snyder et al. 1975). BiOCl is used as a pearlescent white coloring matter in lipsticks (also powder and nail enamel) (Stokinger 1981). Certain over-the-counter drugs sold for gastrointestinal disburbances contain bismuth compounds (e.g., Pepto-Bismol).

B. Inhalation

Ambient air in the United States contains < 0.002 to 0.03 µg Bi/m³ (Snyder et al. 1975). There is a threshold limit value recommended by ACGIH (1983) for Bi_2Te_3 of 10 mg/m³ as a time-weighted average and 20 mg/m³ for a short-term exposure limit. The corresponding values for the Se-doped compound are 5 and 10 mg/m³.

C. Dermal

No information.

D. Total Body Burden and Balance Information

The body burden of Bi in soft tissues of 70-kg reference man is < 230 µg. The estimated bismuth balance is:

Intake, µg/day		Losses, µg/day	
Food and fluids	20	Urine	1.6
Air	< 0.01	Feces	18

III. TOXICOKINETICS

A. Absorption

Soluble bismuth preparations are absorbed since they are detectable in the urine rapidly after intramuscular injections of bismuth solutions. When ingested, insoluble bismuth salts appear to pass unabsorbed and are excreted in feces (Stokinger 1981). Oral absorption of bismuth is estimated to be ∿ 8% (Snyder et al. 1975).

B. Distribution

1. General

In autopsies of treated patients and in dogs given i.m. injections 48-h before, highest concentrations of bismuth were found in kidneys, with the liver next but with considerably smaller concentrations (higher after chronic dosing). About 7.4% of the administered bismuth was retained (Stokinger 1981).

2. Blood

Snyder et al. (1975) estimate < 62 µg Bi in whole blood with < 310 µg in plasma.

C. Excretion

Half of the bismuth retained by the dogs mentioned above was excreted in the first 3 wk, the rest was held for long periods. Bismuth is excreted through the urine, feces, and a small amount in milk (Sallmon 1963; cited by Stokinger 1981).

IV. EFFECTS

A. Acute and Other Short-Term Exposures

Therapeutic use of soluble bismuth compounds have caused signs of toxicity beginning when the total dose reached 2 to 2.5 g or less. Signs begin with foul breath, a blue-black line on the gums, stomatitis, progressing to the ulcerative stage if bismuth intake is continued, as well as general malaise, nausea, weight loss, depression, etc., and skin reactions including serious dermatitis. Death from poisoning occurs from nephritis (Sollman 1963; cited by Stokinger 1981).

Acute toxicity tests in small laboratory animals using soluble bismuth pharmaceuticals results in LD_{50} values ranging from 13 to 82 mg/kg with the median between 21 and 29 mg/kg. The rat oral LD_{50} for BiOCl used extensively in cosmetics was 21.5 g/kg (Stokinger 1981).

B. Chronic Exposure

Humans exposed to Bi_2Te_3 have "tellurium breath," but no adverse health effects. Mild and reversible granulomatous lesions but not fibrosis were seen in laboratory animals exposed to daily inhalations of a mixture of Bi_2Te_3 doped with Bi_2Se_3, SnTe, and Te for 1 y. When exposed to Bi_2Te_3 alone, only "inert" dust responses were seen in the lungs (Stokinger 1981).

C. Biochemistry

1. Effects on Enzymes

No information.

2. Metabolism

No information.

3. Antagonisms and Synergisms

Bismuth can replace lead at its absorption sites putting enough lead in circulation to cause symptoms of lead poisoning (Stokinger 1981).

D. Specific Organs and Systems

In a chronic inhalation study of animals, lung lesions and abnormal lymph nodes were reported. In humans, acute toxic doses affected the stomach, skin, liver, and kidneys with death reported to be caused by nephritis (Stokinger 1981).

E. Teratogenicity

Chick embryos receiving 20 mg $BiCl_3$ on day 4 or 8 via the yolk sac had no abnormalities (Ridgway and Karnofsky 1952; cited by Shepard 1980).

One of four lamb fetuses whose mothers were fed 5 mg Bi/kg daily for extended periods during gestation was stunted, hairless, and exophthalmic (James et al. 1966; cited by Shepard 1980).

F.   Mutagenicity

No information.

G.   Carcinogencity

Rats fed 1, 2, and 5% BiOCl for 2 y showed no excess tumors (Stokinger 1981).   Studies reported no tumors from Bi subcarbonate, Bi dextran, and Bi dimethyl tricarbonate (Stokinger 1981).   A 1-y inhalation study of dogs, rats, and rabbits exposed to 15 mg/m^3 of a Bi_2Te_3, Bi_2Se_3, and SnTe mixture found reversible lung lesions and abnormal lymph nodes, but no evidence that bismuth is carcinogenic (Stokinger 1981).

V.   REFERENCES

ACGIH.   1983.   TLVs$^®$ Threshold limit values for chemical substances and physical agents in the work environment with intended changes for 1983-84. Cincinnati, Ohio: American Conference of Governmental Industrial Hygienists.

Browning E.   1969.   Toxicity of Industrial Metals.   2nd ed.   London: Butterworths.

Snyder WS, Cook MJ, Nasset ES, Karhausen LR, Howells GP, Tipton IH.   1975. International Commission on Radiological Protection.   Report of the task group on reference man.   New York.   ICRP Publication 23.

Shepard TH.   1980.   Catalog of teratogenic agents, 3rd ed.   Baltimore: The Johns Hopkins University Press.

Stokinger HE.   1981.   Chapter 29.   The metals.   In: Patty's industrial hygiene and toxicology.   3rd ed.   Volume 2A.   Clayton GD, Clayton FE, eds. New York: A Wiley-Interscience Publication.   John Wiley and Sons, pp. 1493-2060.

BORON

MAMMALIAN TOXICITY SUMMARY

I. INTRODUCTION

A. Occurrence and Production

Boron is produced from bedded deposits and lake brines in California and Turkey, from a mixture of complex minerals. The element is prepared by reduction by active metals, heat, or electrolytic means (Stokinger 1981).

B. Uses

Major uses include glass (including thermal shock resistant and fiberglass), ceramics, soaps and cleansers, agriculture, and herbicides. Minor uses are legion (Stokinger 1981).

C. Chemical and Physical Properties

Boron is a metalloid with a unique chemistry. Commercial compounds include borates, boron oxides and halides, borohydrides, and boranes (Stokinger 1981).

II. EXPOSURE AND EXPOSURE LIMITS

A. Oral

Boric acid and sodium borate have been used as a food preservative. Boron is widely distributed in foods since it is an essential nutrient for plants. Average drinking water concentrations are 0.114 ppm (Snyder et al. 1975).

B. Inhalation

No data.

C. Dermal

Boric acid is used therapeutically in humans for dermal application as an astringent and antiseptic (Merck Index 1983).

D. Total Body Burden and Balance Information

According to Snyder et al. (1975), the boron balance for the 70-kg reference man in mg/day is: intake from food and fluids 1.3; losses 1.0 in urine, 0.27 in feces, and negligible by other routes. Other estimates of daily boron intake range from 9 to 20 mg/day. The body burden of reference man is estimated to be 200 mg with 140 mg in soft tissues.

III. TOXICOKINETICS

A. Absorption

Boric acid (H_3BO_3) is readily absorbed through the skin, especially of infants (Stokinger 1981). The boron of food and $NaBO_3$ or HBO_3 are ~ 100% absorbed from the gastrointestinal tract (Snyder et al. 1975).

B. Distribution

Some boron is laid down in the skeleton. Data are minimal (Stokinger 1981). Snyder et al. (1975) estimate that 1.1 mg B of a total 200 mg B body burden is in adipose tissue (0.07 µg/g), 0.52 mg in whole blood (0.1 mg/L), 0.1 mg in hair, 0.16 mg in liver, 0.13 mg in lung, 8.1 mg in muscle, 7.4 mg in skeleton, and 1 mg in the epidermis. All other tissues were estimated to have at least an order of magnitude less.

C. Excretion

Ingested boron is excreted primarily in the urine (Snyder et al. 1975).

IV. EFFECTS

A. Acute and Other Short-Term Exposures

Boric acid, whether from ingestion or skin absorption, causes nausea, vomiting, diarrhea, abdominal cramps, and erythematous lesions on skin and mucous membranes. High doses cause circulatory collapse, tachycardia, cyanosis, delirium, convulsion, and coma.

Most other data are from animal studies of the boranes (boron hydrides), exotic compounds formerly considered for use as rocket fuels, now laboratory curiosities (Stokinger 1981).

B. Chronic Exposure

Chronic use of boric acid may cause dry skin, eruptions, and gastric disturbances. Some workers in one boron trifluoride (BF_3) plant had decreased pulmonary function; animal studies showed pulmonary lesions (Stokinger 1981).

C. Biochemistry

1. Effects on Enzymes

No data.

2. Metabolism

No data.

3. Antagonisms and Synergisms

No data.

4. Physiological Requirements

Boron is essential for plants (Snyder et al. 1975).

D. Specific Organs and Systems

Gastrointestinal and dermal symptoms are described in Part IV.A. Lung effects are mentioned in Part IV.B.

E. Teratogenicity

No data.

F. Mutagenicity

No data.

G. Carcinogenicity

No data.

V. REFERENCES

Merck Index, 10th ed. 1983. Rahway, New Jersey: Merck & Co., Inc.

Snyder WS, Cook MJ, Nasset ES, Karhausen LR, Howells GP, Tipton IH. 1975. International Commission on Radiological Protection. Report of the task group on reference man. New York. ICRP Publication 23.

Stokinger HE. 1981. Chapter 29. The metals. In: Patty's industrial hygiene and toxicology. 3rd ed. Volume 2A. Clayton GD, Clayton FE, eds. New York: A Wiley-Interscience Publication. John Wiley and Sons, pp. 1493-2060.

CADMIUM

MAMMALIAN TOXICITY SUMMARY

I. INTRODUCTION

A. Occurrence and Production

Cadmium is found as greenockite (CdS), but it is produced commercially as a by-product of production of other metals (zinc, lead, and copper) from sulfide ores (NIOSH 1976; Stokinger 1981).

B. Uses

The major use of cadmium is electroplating metals to inhibit corrosion. Some compounds (CdS and CdSeS, especially) are used in pigments, primarily for plastics, and as plastic heat stabilizers. A number of alloys are used for soldering, brazing, electrical contacts, and other purposes. Cadmium is also used in Ni-Cd batteries, fungicides, and other minor uses (NIOSH 1976; Stokinger 1981).

C. Chemistry

Cadmium is a soft ductile metal, melting at 321°C. Its usual valence is +2, with chemistry similar to that of zinc. A few compounds have univalent cadmium. The Weston cell, which gives a practically constant voltage of 1.0186 V, contains $CdSO_4$ (Stokinger 1981).

II. EXPOSURE AND EXPOSURE LIMITS

A. Oral

The literature is enormous. A recent study in Europe (Hutton 1983) concluded that the major source of land cadmium is waste disposal; this is greater than the next four sources (coal combustion, iron and steel production, phosphate fertilizers, and zinc production) combined. Cadmium going into water is not as well characterized, but seems to originate mostly from the manufacture of cadmium-containing articles, followed by phosphate fertilizer manufacture, and zinc production. Various national estimates are that the per capita consumption of cadmium varies from 17 to 64 µg/day (USEPA 1980).

In a study of 85 U.S. cities, finished-water levels of cadmium ranged from 0.2 to 0.4 µg/L (US EPA 1975; cited by NAS 1977). In a survey of 380 finished waters in the U.S., cadmium was detected in 0.2% with a mean concentration of 12 µg/L (Kopp 1970; cited by NAS 1977). Another survey of 2,595 water samples detected cadmium 0.1% of of the time with a maximum concentration of 3.94 µg/L (McCabe et al. 1970; cited by NAS 1977). In cities with more acidic water, cadmium levels increased at the tap compared to levels in the finished water; 13% showed an increase in Boston; 51% of samples showed an increase in Seattle with 7% of Seattle's samples exceeding 10 µg/L (Craun and McCabe 1975; cited by NAS 1977). The national interim primary drinking water standard for cadmium is 10 µg/L (USEPA 1975), which is the same as the 1980 criterion (USEPA 1980).

B.   Inhalation

The European study cited above (Hutton 1983) noted that atmospheric cadmium comes from the steel industry and waste incineration, followed by volcanic activity and zinc production. In a U.S. air sampling study, most samples were < 10 ng/m^3, the detection limit. Tobacco smoke contains considerable cadmium, up to 0.1 µg per cigarette in the mainstream and 0.4 to 0.7 µg per cigarette in the sidestream smoke (NIOSH 1976).

The threshold limit value as a time-weighted average in workroom air for an 8-h day is 0.05 mg/m^3 for cadmium dust and salts; the short-term exposure limit is 0.2 mg/m^3 (ACGIH 1983). The OSHA permissible exposure limit is 0.2 mg/m^3 for cadmium dust and 0.1 mg/m^3 for cadmium fumes (OSHA 1981).

C.   Dermal

There is no evidence of dermal absorption.

D.   Total Body Burden and Balance Information

According to Snyder et al. (1975), the cadmium balance for the 70-kg reference man (in µg/day) is: intake 150 from food and fluids and < 1 from air; losses 100 in urine and 50 in feces. The body burden is estimated at 50 mg (38 mg in soft tissues). The burden is about doubled for a smoker (Lauwerys 1983).

III. TOXICOKINETICS

A.   Absorption

Oral absorption of environmental cadmium is poor: only about 5%. However, up to 50% of respired cadmium is absorbed (USEPA 1980; Friberg et al. 1977).

B.   Distribution

1.   General

Absorbed cadmium is concentrated in the liver and kidneys, apparently in association with metallothionein, a cadmium-binding protein (USEPA 1980; Stokinger 1981; Friberg et al. 1977). Kjellstroem et al. (1984) recently proposed a new measure of the critical concentration of cadmium in kidney cortex: the population critical concentration (PCC) with a clearly defined response rate. The 10% response (PCC-10) is 180 to 220 mg/kg.

2. Blood

Snyder et al. (1975) estimated 36 µg in the whole blood of 70-kg reference man.   More than 70% of the cadmium in the blood is bound to the erythrocytes.   Levels in whole blood of nonoccupationally exposed persons are generally < 10 µg/L (median 1.5 µg/L).

Roels et al. (1981) found that blood levels of cadmium correlated to recent exposures, while liver and renal cortex levels, as well as urine concentrations, correlated to body burden, i.e., chronic exposures.

Several recent studies by Lauwerys et al., Buchet et al., and Roels et al. to evaluate critical cadmium concentrations in blood, urine, and renal cortex were summarized by Lauwerys (1983). [Some of these are cited in the references in this profile.   The neutron activation analysis methodology of these was good.]   Signs of renal dysfunction are greater than in the control group when there is > 10 µg Cd/G creatinine in the urine (corresponding to 160 ppm Cd in the renal cortex).   Their data suggest 10 µg Cd/L whole blood would be a no-adverse-effect level for chronic cadmium exposure.

DeSilva and Donnan (1981) found that for cadmium pigment exposed workers, blood cadmium concentrations did not correlate well with years of exposure or with cadmium in urine.   The blood concentrations did correlate reasonably well with degree of exposure after ~ 2 yr.   Thus, two of three workers exposed to total dust concentrations 0.7 mg/m³ and respirable dust 0.3 mg/m³ had no or slight respiratory abnormalities and proteinuria when blood concentrations were 18 and 24 µg/L, but the other worker had mild airways obstruction, bronchitis, and renal tubular damage; his blood concentration was 35 µg/L.   Three workers exposed to total dust at 1.5 to 2 mg/m³ and respirable dust at 0.7 mg/m³ suffered from dyspnea and moderate or severe airways obstruction.   Two also had renal damage.   Blood concentrations were 44, 54, and 73 µg Cd/L.   [The analytical methodology of this paper was excellent; the major shortcoming was that few subjects were used.]

Murray et al. (1981), using poorly described analytical methodology [Roels et al. 1978, were cited], concluded that although they reflect acute exposures, blood and urine levels cannot be used for predicting possible nephropathy.   For example, cadmium in blood of exposed workers did not correlate with any biochemical parameter that would indicate potential renal dysfunction.   They suggested that hair cadmium is a good measure of exposure.

Baddeley et al. (1983) found blood cadmium levels > 10 µg/L in 6 of 57 exposed workers (10.5%) and 2 of 54 unexposed persons (3.7%).   Raised concentrations were more common in liver (27.7% of exposed persons had > 20 ppm vs. 3.2% of nonexposed persons) and urine (43.1% of exposed workers had > 2 µg Cd/L vs. 7.8%).   Moreover, raised blood concentrations correlated poorly with raised liver and raised urine concentrations.   (The neutron activation analysis methodology was good.)

3.   Adipose

Snyder et al. (1975) estimated 0.7 mg Cd (0.05 μg/g) in adipose tissue of reference man.

C.   Excretion

Cadmium is excreted in the urine and, via the bile, in the feces. Small quantities come out in the hair (USEPA 1980; Stokinger 1981). A time lag is seen in newly exposed workers before urinary cadmium correlates with exposure. By the time renal dysfunction is present, cadmium loss increases (Lauwerys 1983).

IV. EFFECTS

A.   Acute and Other Short-Term Exposures

Acutely, cadmium is a powerful emetic. The effects of ingestion include persistent vomiting (beginning only 15 to 30 min after ingestion), increased salivation, choking sensations, abdominal pain, tenesmus, and diarrhea.

Acute inhalation toxicity is characterized by delayed (4 to 10 hr) dyspnea, cough and tightness in the chest. This develops to pulmonary edema and possibly bronchopneumonia (USEPA 1980; Stokinger 1981).

B.   Chronic Exposure

Chronic exposure produces a wide variety of effects, which are discussed separately below. Affected organs include the kidney, lung, heart, bones, and gonads (USEPA 1980; Stokinger 1981).

C.   Biochemistry

1.   Effects on Enzymes

Cadmium induces changes in several gluconeogenic enzymes. It reduces serum antitrypsin concentration and depresses the trypsin inhibitor capacity. Other effects of cadmium include alteration of oxidative phosphorylation and causation of enzymuria (Stokinger 1981).

2.   Metabolism

Cadmium is known to bind many proteins, particularly the sulfate and carbonyl groups. This is commonly seen as morphological alteration of various subcellular structures (Stokinger 1981). It is bound mainly to metallothionein in all tissues (Lauwerys 1983).

3.   Antagonisms and Synergisms

Cadmium toxicity is decreased by the presence of other metal ions, especially zinc, although calcium, copper, and iron also interact. Selenium antagonizes cadmium toxicity as well (USEPA 1980; Stokinger 1981).

D.   Specific Organs and Systems

1.   Kidney

The kidney is the organ of greatest toxicological interest after chronic cadmium poisoning. It is generally accepted that an accumulation of more than 200 mg/kg wet weight in the renal cortex leads to frank toxicity; this corresponds to a urinary excretion of 10 mg Cd/kg creatinine, under chronic (steady-state) conditions (Friberg et al. 1977; USEPA 1980). A recent study (Buchet et al. 1981) confirmed this threshold (see also part III.B.1.). Other organs are affected at doses higher than those that affect the kidney.

Since cadmium first affects the proximal tubule's reabsorption capabilities, the first effect to be detected is proteinuria. Later signs include amino aciduria, glucosuria, decreased urine-concentrating ability, and abnormalities in handling uric acid, calcium, and phosphorus. The mineral problems may lead to kidney stones and osteomalacia. It is not known if the kidney damage is reversed on cessation of exposure (Friberg et al. 1977; NIOSH 1976; Stokinger 1981).

2.   Lung

Chronic cadmium inhalation produces pulmonary emphysema (NIOSH 1976; Stokinger 1981).

3.   Heart and Cardiovascular System

Anemia has been seen after many years of cadmium exposure. This seems to be due to deficient iron absorption from food, rather than bone marrow toxicity (Friberg et al. 1977; Stokinger 1981).

No definite relationship between cadmium and cardiovascular disease, especially hypertension, has been demonstrated, but the subject remains highly controversial (NIOSH 1976; Stokinger 1981). This subject has been reviewed recently (Shaper 1979; Revis 1982).

4.   Skeleton

A variety of skeletal effects, including back and extremity pain, difficulty in walking, pseudofractures and osteomalacia (softening of the bone due to mineral loss), are probably secondary to the derangement of mineral metabolism caused by the kidney effects. Other factors, such as nutritional deficiencies, influence the degree of the symptoms. This is well illustrated by the Japanese itai-itai ("ouch-ouch") disease, induced by environmental cadmium primarily in post-menopausal women (low estrogen levels) with deficient diets (Friberg et al. 1977; NIOSH 1976; Stokinger 1981).

5. Gonads

Cadmium toxicity includes suppression of testicular function in some cases (NIOSH 1976; Stokinger 1981).

6. Olfactory Organs

Cadmium fumes can damage the olfactory organs. Some cases progress to total anosmia (NIOSH 1976; Stokinger 1981).

E. Teratogenicity

Cadmium was teratogenic in the few animal studies reported. One human epidemiological study, not of the finest quality, reported rickets. It has been hypothesized that these effects are due to zinc deficiency in the fetus due to cadmium-induced zinc deficiency in the mother (Friberg et al. 1977).

In pregnant women afflicted by itai-itai disease, no fetal defects were observed (Shepard 1980).

F. Mutagenicity

Limited data show highly inconsistent results (Friberg et al. 1977; Stokinger 1981; USEPA 1980).

G. Carcinogenicity

Cadmium may cause prostrate cancers in man, but the evidence is not yet conclusive (Friberg et al. 1977; NIOSH 1976; Stokinger 1981).

V. REFERENCES

ACGIH. 1983. TLVs® Threshold limit values for chemical substances and physical agents in the work environment with intended changes for 1983-84. Cincinnati, Ohio: American Conference of Governmental Industrial Hygienists.

Baddeley H, Thomas BJ, Thomas BW, Summers V. 1983. Liver cadmium concentrations in metal industry workers. Br J Radiol 56(667):449-451.

Buchet JP, Roels H, Lauwerys R, Bruaux P, Claeys-Thoreau F, Lafontaine A, Verduyn G. 1980. Repeated surveillance in exposure to cadmium, manganese, and arsenic in school-age children living in rural, urban, and nonferrous smelter areas in Belgium. Environ Res 22:95-108.

Buchet JP, Roels H, Bernard A Jr., Lauwerys R. 1981. Assessment of renal function in workers simultaneously exposed to inorganic lead and cadmium. J Occup Med 23(5):348-352.

DaSilva PE, Donnan MB.   1981.   Chronic cadmium poisoning in a pigment manufacturing plant.   Br J Ind Med 38:76-86.

Friberg L, Nordberg G, Piscator M.   1977.   Cadmium.   In: Toxicology of metals-volume II.   Springfield, Virginia: National Technical Information Service, pp. 124-163.

Hutton M.   1983.   Sources of cadmium in the environment.   Ecotoxic Environ Saf 7:9-24.

Kjellstrom T, Elinder CG, Friberg L.   1984.   Conceptual problems in establishing the critical concentration of cadmium in human kidney cortex. Environ Res 33(2):284-295; Chem Abstr 1984.   100:214867v.

Lauwerys R, Roels H, Bernard A, Buchet AP.   1980.   Renal response to cadmium in a population living in a nonferrous smelter area in Belgium.   Int Arch Occup Environ Health 45:271-274.

Lauwerys RR.   1983.   Chapter II.   Biomedical monitoring of exposure to inorganic and organometallic substances.   In: Industrial chemical exposure: Guidelines for biological monitoring.   Davis, CA: Biomedical Publications, pp. 9-50.

Murray T, Walker BR, Spratt DM, Chappelka R.   1981.   Cadmium nephrotoxicity: Monitoring for early evidence of renal dysfunction.   Arch Environ Health 36(4):165-171.

NAS.   1977.   Natl. Academy of Sciences.   Drinking water and health.   Vol. I. Washington, DC.

NIOSH.   1976.   Natl. Inst. Occupational Safety and Health.   Criteria for a recommended standard...occupational exposure to cadmium.   Washington, DC: U.S. Government Printing Office, HEW Publication No. (NIOSH) 76-192.

OSHA.   1981.   Occupational Safety and Health Admin.   Occupational safety and health standards.   Subpart 2--Toxic and hazardous substances.   Code of federal regulations 29 (Part 1910.1000) pp. 673679.

Revis NW.   1982.   Relationship of vanadium, cadmium, lead, nickel, cobalt and soft water to myocardial and vascular toxicity and cardiovascular disease. In: Cardiovascular toxicology.   Van Stee EH, Ed.   New York, NY: Raven, pp. 365-375.

Roels HA, Lauwerys RR, Buchet JP, Bernard A, Chettle DR, Harvey TC, Al-Haddad IK.   1981.   In vivo measurement of liver and kidney cadmium in workers exposed to this metal: Its significance with respect to cadmium in blood and urine.   Environ Res 26:217-240.

Shaper AG.   1979.   Cardiovascular disease and trace metals.   Proc Royal Soc London.   B205:135-143.

Shepard TH.   1980.   Catalog of teratogenic agents, 3rd ed.   Baltimore: The Johns Hopkins University Press.

Snyder WS, Cook MJ, Nasset ES, Karhausen LR, Howells GP, Tipton IH.   1975. International Commission on Radiological Protection.   Report of the task group on reference man.   New York.   ICRP Publication 23.

Stokinger HE.   1981.   Chapter 29.   The metals: In: Patty's industrial hygiene and toxicology, 3rd ed., volume IIA, Toxicology.   Clayton GD, Clayton FE, eds.   New York: Wiley-Interscience, John Wiley & Sons, pp. 1493-2060.

USEPA 1980.   Environmental Protection Agency.   Ambient water quality criteria for cadmium.   Springfield, VA: National Technical Information Service, PB81-117368.

CALCIUM

MAMMALIAN TOXICITY SUMMARY

I. INTRODUCTION

A. Occurrence and Production

Calcium metal is always found combined in compounds in nature. The principal commercial source of calcium is limestone. Calcium is produced by electrolysis of $CaCl_2$ and by the thermal reduction of limestone by Al or Si. Calcium chloride is recovered from natural salt brines and is a by-product of the ammonia soda (Solvay) process (Merck Index 1983).

B. Uses

Calcium metal is used in metallurgy as a deoxidizer for Cu, Be, and steel (with Si), to harden lead for bearings, and to alloy Ce for flints. Principal uses of $CaCl_2$ are for melting ice and snow, to accelerate concrete setting, and to settle dusts on unpaved roads. Numerous calcium compounds have therapeutic uses (Merck Index 1983). For example, calcium salts are administered in all cases of low-calcium tetany. They are also used in antispasmodics, diuretics, and antacid preparations and for their circulatory actions (Arena 1973).

C. Chemical and Physical Properties

Calcium, atomic no. 20, is an alkaline earth metal. Its compounds exhibit +2 valence. Calcium melts at 850°C and boils at 1440°C and burns when exposed to air.

II. EXPOSURE AND EXPOSURE LIMITS

A. Oral

Estimates of calcium in the diet range from 690 to 1,130 mg/day (Snyder et al. 1975). Milk and cheese are the richest dietary sources (NAS 1980).

B. Inhalation

No information.

C. Dermal

No information.

D. Total Body Burden and Balance Information

The calcium balance for 70-kg reference man, according to Snyder et al. (1975), is as follows:

Intake, mg/day	Losses, mg/day	
Food and fluids 1,100	Urine	180
	Feces	740
	Sweat	32-150
	Other fluids	Trace
	Hair	Trace

A 70-kg adult contains ∿ 1,200 g calcium (NAS 1980).

III. TOXICOKINETICS

A. Absorption

Calcium is incompletely absorbed from the gastrointestinal tract; normally 70 to 80% of the intake is excreted in the feces. Calcium absorption may decrease with advancing age. Efficient absorption requires Vitamin D. Oxalates and phytates in the diet form insoluble calcium salts; but if calcium intake is liberal, the effect on absorption is not important. A Ca:P ratio of 2:1 may maximize Ca absorption and minimize bone loss, but the current recommended daily allowances for Ca and P give a ratio of 1:1 (NAS 1980).

B. Distribution

1. General

About 99% of the total body burden (∿ 1,200 g) of a 70-kg adult is contained in the bones. About 700 mg enters and leaves the bones daily. Soft tissues and extracellular fluids contain ≦ 10 g Ca (NAS 1980).

2. Blood

The daily variation in Ca concentration in the plasma is ± 3%. Homeostatic mechanisms involve hormonal control of hypocalcemia and hypercalcemia. Very high calcium concentrations in the serum or urine may accompany certain diseases (NAS 1980). Serum levels > 160 mg/L may be fatal (Arena 1973). When plasma Ca falls from the normal 100 mg/L to 70 mg/L, tetany (muscular twitching) develops. At 35 to 50 mg/L plasma, convulsions occur (Harrow and Mazur 1962).

C. Excretion

Unabsorbed calcium is excreted in the feces. Even major increases in calcium intake affect urinary excretion only slightly. There is wide individual variation in calcium urinary excretion. Increases in purified protein intake increase calcium excretion significantly, but this does not occur on a high-meat diet (NAS 1980).

IV. EFFECTS

A. Acute and Other Short-Term Exposures

Repeated i.v. injections of Ca gluconate (9% Ca) have caused a sensation of heat, fever, nausea, vomiting, and oppression in the chest (Arena 1973). See also Part IV.D for irritant effects of $CaCl_2$ solutions.

B. Chronic Exposure

Hypercalcemia (elevated blood calcium concentrations) occurs in diseases such as hyperparathyroidism, sarcoidosis, malignancy, and vitamin D poisoning. Sudden death may occur if calcium levels remain above 160 mg/L. Calcium toxicity signs and symptoms include anorexia, nausea, vomiting, dehydration, lethargy, coma, and death.

Hypercalcemia may cause pathological kidney changes and abnormal calcium salt deposits (calcification) in soft tissues (Harrow and Mazur 1962). Idiopathic hypercalcemia is associated with vitamin D intoxification in infants. The bones become denser, mental deterioration progresses to idiocy, and calcium deposits in the kidneys lead to chronic uremia (Dorland 1974). High calcium intakes, however, are not the primary cause of hypercalcemic diseases (NAS 1980).

C. Biochemistry

1. Effects on Enzymes and Normal Functions

Enzyme involvement with calcium was not mentioned, but calcium is essential for blood coagulation, myocardial function, muscle contractility, and the integrity of intracellular cement substances and various membranes (NAS 1980).

2. Metabolism

Most calcium is held in the skeleton as crystallized Ca phosphates within an organic matrix. Amorphous Ca phosphates predominate in early life, but they are eventually converted to the crystalline form (NAS 1980).

3. Antagonisms and Synergisms

EDTA, oxalate, phytate, and phosphate may affect calcium absorption or blood levels. For example, excessive phosphorus intake may lead to increased bone resorption and increased fecal calcium loss. Oxalate is added to blood in vitro to bind Ca^{2+} and prevent clotting. Sulfate will reduce serum Ca^{2+} by forming an ion-pair that is not reabsorbed (NAS 1980; Arena 1973).

4. Physiological Requirements

Because of high dietary levels of protein and phosphate, NAS (1980) recommended 800 mg as the daily allowance for adults. The recommendation was 1,200 mg/day for lactating and pregnant women, 60 mg/kg bw for infants, 800 mg/day for children 1 to 10, and 1,200 mg/day for children 10 to 18.

D. Specific Organs and Systems

1. Skin

A 27% $CaCl_2$ solution causes erythema, exfoliation, ulceration, and scarring of the skin. Injections into the tissues should not be given because perivenous infiltration and necrosis may occur (Arena 1973).

2. Gastrointestinal Tract

Without a demulcent vehicle, oral doses of $CaCl_2$ solutions will cause irritation and ulceration of the gastrointestinal tract. Gastrointestinal distress occurs in calcium intoxication (Arena 1973).

3. Kidney

Kidney damage and kidney stones may develop in hypercalcemia (Harrow and Mazur 1962; NAS 1980).

E. Teratogenicity

Hypercalcemia has been associated with congenital heart disease (supravalvular aortic stenosis) (Black and Bonham-Carter 1963; cited by Shepard 1980).

F. Mutagenicity

No information.

G. Carcinogenicity

No information.

V. REFERENCES

Arena JM. 1973. Poisoning: toxicology--symptoms--treatments. Springfield, IL: Charles C. Thomas Publisher.

Dorland. 1974. Dorland's illustrated medical dictionary, 25th ed. Philadelphia, PA: W. B. Saunders.

Harrow B, Mazur A. 1962. Textbook of biochemistry, 8th ed. Philadelphia, PA: W. B. Saunders Co.

Merck Index, 10th ed. 1983. Rahway, New Jersey: Merck & Co., Inc.

NAS. 1980. National Academy of Sciences. Recommended dietary allowances. 9th ed. Washington, DC: Pringing and Publishing Office, National Academy of Sciences.

Shepard TH. 1980. Catalog of teratogenic agents, 3rd ed. Baltimore, MD: The Johns Hopkins University Press.

Snyder WS, Cook MJ, Nasset ES, Karhausen LR, Howells GP, Tipton IH. 1975. International Commission on Radiological Protection. Report of the task group on reference man. New York. ICRP Publication 23.

CESIUM

MAMMALIAN TOXICITY SUMMARY

I. INTRODUCTION

A. Occurrence and Production

Cesium occurs in nature as pollucite (Cs-Al-Na silicate), carnallite, beryl, and other minerals (Browning 1969) including the aluminosilicate lepidolite and the borate rhodizite. Its average crustal abundance is 1 ppm (Merck Index 1983). Mineral resources in the United States are in Maine and South Dakota (Stokinger 1981).

B. Uses

Cesium is used in photoelectric cells and as a polymerization catalyst. The radioisotope ^{137}Cs has been used in the treatment of skin disorders, such as cutaneous epithelioma, and deep thoracic tumors (Browning 1969). Its compounds are used, for example, in x-ray fluorescent screens (CsBr, CsCl, CsI), as a getter in vacuum tubes (CsCl), as catalysts (Cs_2Co_3, CsOH, Cs_2SO_4), as spectrometer prisms (CsBr, CsI), and in scintillation counters (Merck Index 1983).

C. Chemical and Physical Properties

Cesium, atomic no. 55, is an alkali metal with valence +1 in its common compounds. It melts at 28.5°C and boils at 705°C. Many common compounds are water soluble. Cs perchlorate, permanganates, fluorosilicates, and periodates are relatively water insoluble (Merck Index 1983). In its recovery, it is precipitated as $2CsCl \cdot PbCl_4$ and Cs Zn ferrocyanide. CsOH is the strongest base known and must be stored in silver or platinum. Cesium forms three oxides Cs_2O, Cs_2O_2, and Cs_2O_3 (Stokinger 1981).

II. EXPOSURE AND EXPOSURE LIMITS

A. Oral

Almost half of the $\sim$ 10 μg Cs/day from the diet in Japan is contributed by meat, eggs, and milk products. North American diets could contribute more cesium (Snyder et al. 1975). ^{137}Cs tissue concentrations in tissues and food sources increased after nuclear test explosions. For ^{137}Cs the permissible annual radiation dose is 100 mrem (Stokinger 1981).

B. Inhalation

The amount of cesium inhaled per day is estimated as 0.025% of the amount ingested (Snyder et al. 1975). The threshold limit value for CsOH in workplace air is 2 mg/m³ as a time weighted average for an 8-h work day (ACGIH 1983). Other cesium compounds are considered in the nuisance dust category (10 mg/m³) (Stokinger 1981).

C. Dermal

No information.

D. Total Body Burden and Balance Information

Snyder et al. (1975) estimates a body burden of 1.5 mg and gives the following balance information for cesium for 70-kg reference man:

Intake, μg/day		Losses, μg/day	
Food and fluids	10	Urine	9.0
Airborne	0.025	Feces	4.0
		Sweat, Saliva	?

III. TOXICOKINETICS

A. Absorption

Oral doses of ^{137}Cs were rapidly and almost completely absorbed (Snyder et al. 1975).

B. Distribution

1. General

In potassium deficiency followed by feeding cesium and potassium, rat skeletal muscles showed a final intracellular/extracellular ratio for cesium that was ∿ 4.5 times greater than the same ratio for potassium. Like potassium and rubidium, cesium is contained much more in red cells, muscle tissue, and viscera than in bone and plasma (Browning 1969). In distribution studies, Cs was retained in the muscle. During the first week, however, accumulation was greater in liver, spleen, and kidney. Cesium does not cross the placenta. Skeleton was one of the lowest repositories (Stokinger 1981).

2. Blood

Cesium concentrates in the red cells like potassium but is exchanged more slowly (Browning 1969). In the distribution study mentioned above, blood concentrations of 17 tissues were always lowest (Stokinger 1981). Snyder et al. (1975) give 15 μg as the total cesium content in whole blood.

3. Adipose

Snyder et al. (1975) give no estimate of adipose tissue content of cesium.

C. Excretion

The major excretory route is in the urine. Seventy percent of an intravenous dose of ^{137}Cs was excreted within 7 days; 12 times as much was in the urine as in the feces. Increased dietary potassium increases cesium excretion (Browning 1969). The biologic half-life of ^{137}Cs in two cancer patients was 50 to 60 days (Stokinger 1981).

IV. EFFECTS

A. Acute and Other Short-Term Exposures

^{137}Cs given internally to a human in a single dose at 60 to 106 µCi produced a liver disorder and a neuroendocrine disturbance (Browning 1969).

CsOH had an i.p. LD_{50} in rats of 89 mg/kg. Its high toxicity may have been due to its alkalinity. CsBr and CsI were less toxic (LD_{50}'s of 874 and 715 mg/kg, respectively). CsCl was the least toxic; its i.p. LD_{50} was 1,118 mg/kg (Cochran et al. 1950). No specific histological lesions have been reported. A low-potassium diet containing $\sim$ 0.39% Cs caused animals to become increasingly hyperirritable. They finally died in convulsions (Browning 1969).

In mice, CsOH and CsCl had oral LD_{50}'s of 800 and 2,300 mg/kg, respectively. The oral LD_{50} for CsOH in rats was 1,026 mg/kg; for CsI, 2,386 mg/kg. As would be expected from a strong base, CsOH is extremely irritating to abraded skin and the eye (Stokinger 1981).

Reports of accidental or intentional ^{137}Cs exposure in the radium luminizing industry, in experimentation with ^{137}Cs, or in treatment of malignancies do not indicate that the exposures were great enough to incur harmful effects from the radiation let alone the toxicity of the metal itself. Industrial exposure in the luminizing industry was < 2% than from natural food sources in 1958 and 1959 (contaminated by fallout) (Stokinger 1981).

B. Chronic Exposure

No cases of industrial problems from cesium exposure (Browning 1969) and no chronic toxicity studies in animals have been reported (Stokinger 1981).

C. Biochemistry

1. Effects on Enzymes

CsCl did not inhibit adenosine triphosphatase activity of mouse liver in vitro (Cochran et al. 1950).

2. Metabolism

Cesium is able to replace potassium to some extent. For example, it partially protects kidney and heart in potassium deficiency conditions, it can increase acetylcholine formation in respiring brain tissue, and it concentrates in erythrocytes as does potassium (Browning 1969).

3. Antagonisms and Synergisms

No information.

D. Specific Organs and Systems

No information, but CsOH would be very corrosive to skin, eye, and mucous membrane.

E. Teratogenicity

Radiocesium (^{137}Cs) (2 to 4 µCi/g bw) i.v. injections on day 4, 8, 10, 12, or 15 in pregnant mice gave a small incidence of cleft palates or tail anomalies (Hiratu 1964; cited by Shepard 1980).

F. Mutagenicity

No information.

G. Carcinogenicity

No information.

V. REFERENCES

ACGIH. 1983. TLVs[®] Threshold limit values for chemical substances and physical agents in the work environment with intended changes for 1983-84. Cincinnati, Ohio: American Conference of Governmental Industrial Hygienists.

Browning E. 1969. Toxicity of industrial metals. 2nd ed. London: Butterworths.

Cochran KW, Doull J, Mazur M, DuBois KP. 1950. Acute toxicity of zirconium, columbium, strontium, lanthanum, cesium, tantalum and yttrium. Arch Ind Hyg Occup Med 1:637-650.

Merck Index, 10th ed. 1983. Rahway, New Jersey: Merck & Co., Inc.

Shepard TH. 1980. Catalog of teratogenic agents, 3rd ed. Baltimore, MD: The Johns Hopkins University Press.

Snyder WS, Cook MJ, Nasset ES, Karhausen LR, Howells GP, Tipton IH. 1975. International Commission on Radiological Protection. Report of the task group on reference man. New York. ICRP Publication 23.

Stokinger HE. 1981. Chapter 29. The metals. In: Patty's industrial hygiene and toxicology, 3rd ed., volume IIA, Toxicology. Clayton GD, Clayton FE, eds. New York: Wiley-Interscience, John Wiley & Sons, pp. 1493-2060.

CHROMIUM

MAMMALIAN TOXICITY SUMMARY

I. INTRODUCTION

A. Occurrence and Production

The earth's crust contains an average 125 ppm chromium. Trivalent and hexavalent chromium are both found in nature, but Cr(III) predominates. Chromite, $FeOCr_2O_3$, is the only important chromium ore mineral. Direct reduction of the ore gives ferrochromium. Chemical treatment of high carbon ferrochromium followed by electrolysis gives chromium metal. Chromium may also be produced by reduction of chromium compounds. Roasting chromite ore with soda ash or soda ash and lime gives impure sodium chromate and dichromate, the source of most other chromium compounds (Langard and Norseth 1977).

B. Uses

Chromium compounds are used in tanning, pigments, and electroplating and as catalysts, corrosion inhibitors, and wood preservatives (NAS 1980; Merck Index 1983). Other compounds are used as refractories and foundary sands. Chromium metal is an essential ingredient of stainless steel, superalloys for jet engines, and other alloys (NAS 1974).

C. Chemistry

Chromium, atomic number 24, is a very hard metal melting at 1857 ± 20°C and boiling at 2672°C. Its major oxidation states are 2, 3, and 6. Cr(II) (chromous ion) is rapidly oxidized to the Cr(III) (chromic) form. Cr(VI) exists as chromates, which are strong oxidizing agents (Fowler 1977).

II. EXPOSURE

A. Oral

Foods contain up to ~ 0.5 ppm chromium (wet weight). About 0.03 to 0.1 mg is the estimated daily chromium intake from food, the major environmental source. Meat, vegetables, and unrefined sugar are good sources. Glucose tolerance factor, found predominantly in yeast, liver, and meats, is the most biologically available form of chromium (Langard and Norseth 1977).

The USEPA standard for total chromium in domestic water supplies is 50 µg/L. The USEPA criterion for Cr(III) ingested through water (95%) and contaminated aquatic organisms (5%) is 170 mg/L [sic]. The criterion for Cr(VI) is 50 µg/L (USEPA 1980). Drinking waters have been reported to contain up to 35 µg/L (NAS 1980). Chromium levels in the finished water of the 100 largest U.S. cities were as high as 35 µg/L with median level of 0.43 µg/L (Durfor and Becker 1964; cited by NAS 1977). Another study of 85 U.S.

cities found the level of chromium to be < 5.0 to 6.0 µg/L (USEPA 1975; cited by NAS 1977). Tapwater samples from Dallas, Texas, averaged 4 µg/L with the range of chromium 1 to 20 µg/L (NAS 1977). In cities with more acidic water, chromium levels increased at the tap over levels in the finished water, 13% showed an increase in Boston, and 51% of samples showed an increase in Seattle with 7% of Seattle's samples exceeding 10 µg/L (Craun and McCabe 1975; cited by NAS 1977).

B. Inhalation

Ambient air generally contains < 10 to ∿ 50 ng Cr/m³ (Langard and Norseth 1977; NAS 1980). Chromium concentration in the tissues generally decreases from birth, but increases in the lung from about age 10, indicating deposition from inhaled air (Langard and Norseth 1977).

The threshold limit value as a time-weighted average in workroom air for an 8-h day is 0.5 mg Cr/m³ for chromium metal and Cr(II) and Cr(III) compounds. The short-term exposure limit is 0.05 mg Cr/m³ for water-soluble chromium compounds (ACGIH 1983). The OSHA permissible exposure limit is 0.5 mg Cr/m³ for soluble chromium (OSHA 1981). However, NIOSH (1976) has recommended 0.025 mg/m³ [0.001 mg/m³ for carcinogenic Cr(VI)] as the limit.

C. Dermal

Even persons without occupational exposure to chromium(VI) exhibit dermal hypersensitivity. Besides occupational dermatitis, chromate workers develop ulcerations and perforations of the nasal septum (Langard and Norseth 1977).

D. Total Body Burden and Balance Information

If daily urinary excretion is 2 to 10 µg/L and daily intake is 30 to 100 µg, at least 10% of chromium in the diet is absorbed (Langard and Norseth 1977). However, more recent measurements indicate these values are about an order of magnitude too high (NAS 1980).

Snyder et al. (1975) estimated the following balance information for 70-kg reference man:

Intake, µg/day		Losses, µg/day	
Food and fluids	150	Urine	70
Air	0.1	Feces	80
		Sweat	1
		Hair, nails	0.6
		Other fluids	Trace

III. TOXICOKINETICS

A. Absorption

Chromium may be absorbed from the lungs although at least some is deposited in human lungs in an insoluble form. Less than 1% of an oral dose of Cr(III) is absorbed; about 2% of oral doses of chromates is absorbed in humans. Perhaps no more than 1% of chromium in foods, probably in the form of glucose tolerance factor, is absorbed (Langard and Norseth 1977; NAS 1980).

B. Distribution

1. General

After large doses, the reticuloendothelial system, the liver, the spleen, and the bone marrow of laboratory mammals accumulate inorganic chromium. Highest concentrations from glucose tolerance factor, however, are in the liver, uterus, kidney, and bone. Cr(III) and glucose tolerance factor cross the placenta to the fetus. Cr(III) is the form found in biological tissues (Langard and Norseth 1977; NAS 1974).

Hair values range from 200 to 2,000 µg Cr/kg. Lung values of 140 to 700 µg/kg have been reported with lower concentrations in other organs (Langard and Norseth 1977).

2. Blood

Except for chromates, all chemical forms of chromium are rapidly cleared from the blood. Normally, chromium is distributed equally in the erythrocytes and plasma, but occupational exposure increases the chromium mainly in the red cells (Langard and Norseth 1977). The following blood and urine concentrations reported by the same reference for chromate workers were listed by NAS (1974):

	ng/g or ng/mL
Blood cells	30, 54, 140
Plasma	0, 20, 17
Urine	52, 0, 3.6, 0

Stokinger (1981) concluded that neither blood nor urinary concentrations of chromium are of practical use to indicate exposure or body burden. However, NIOSH (1975) includes urinalysis as part of a worker medical surveillance program. Lauwerys (1983) also stated that blood chromium values are apparently of little or no value in monitoring exposure, but soluble Cr(VI) exposure can be correlated with chromium in urine. See Part III.C.

Versieck (1984) pointed out that determination of elements such as chromium that are present in trace amounts in blood serum involve the risk of introducing errors by sample contamination. Many literature values appear to be grossly erroneous. Versieck et al. (1978; cited by Versieck 1984) determined a mean 0.160 µg Cr/L (range 0.0382 to 0.351 µg/L and standard deviation 0.083 µg/L) in human blood plasma or serum by neutron activation analysis and cited a 1979 report by Nakahara et al., who used the same method but found a mean of 782 µg Cr/L with a standard deviation of 495 µg/L.

C. Excretion

The major excretory route for absorbed chromium is in the urine. About 60% of the filtered amount in the kidneys is reabsorbed by the tubules. Three compartments of chromium in the rat have biological half-lives of 0.5, 5.9, and 83.4 days, respectively (Langård and Norseth 1977).

Workers exposed to Cr(VI) at 0.05 mg/m^3 for 8 h excrete 30 µg Cr/g creatinine, more than six times greater than normal values (Lauwerys 1983).

IV. EFFECTS

A. Acute and Other Short-Term Exposures

Soluble chromates have high oral LD$_{50}$'s (∼ 1,500 mg/kg) and intermediate dermal LD$_{50}$'s (∼ 200 to 350 mg/kg). They are highly toxic (10 to 50 mg/kg) by injection. The insoluble chromates have intermediate LD$_{50}$'s when injected. Cr(II) and Cr(III) have much lower acute toxicities for these routes.

High chromate doses by injection or skin or mucous membrane absorption damage the kidneys of laboratory mammals and humans. Pulmonary hyperemia and inflammation were observed in rabbits inhaling chromic acid dusts. One human case of enlarged, tender liver from acute poisoning by K dichromate has been reported (Stokinger 1981).

B. Chronic Exposure

Long-term inhalation or ingestion of Cr(III) compounds reveals no adverse health effects, but long-term inhalation exposure to insoluble Cr(VI) compounds is associated with lesions of the mucosa and submucosa of the respiratory tract and other toxic effects. Several occupational groups are exposed to chromium(VI) compounds. Signs and symptoms from chronic inhalation exposure include allergic contact dermatitis, skin ulcers, nasal membrane inflammation and ulceration, nasal septum perforation, perforated eardrums, rhinitis, nosebleed, liver damage, pulmonary congestion and edema, epigastric pain, tooth erosion and discoloration; and nephritis. An excessive incidence of lung cancer occurs in workers of the chromate producing industry and possibly of the pigment producing industry. Cr(III) has also been implicated in lung cancer cases. Gastric cancers, presumably from excessive mouth breathing or from swallowing inhaled dusts, have also been reported. Other effects include gastrointestinal ulcers or hypertrophic gastritis, decrease or loss of the sense of smell, and blood changes (leukopenia or leukocytosis, monocytosis, and eosinophilia) (Stokinger 1981; NIOSH 1975).

C. Biochemistry

1. Effects on Enzymes

Phosphoglucomutase requires chromium for its activity. The succinate, cytochrome C reduction system is catalyzed by chromium. Chromium may also inhibit certain enzymes, depending on its concentration (Stokinger 1981).

2. Metabolism

Chromium complexes and precipitates proteins, which probably explains its toxic actions in dermatitis and sensitization, its cancer induction by combination with RNA and DNA, and the formation of chrome ulcers and perforated nasal septums. Chromium also reacts with many substances critical for proper bodily functions. Certain amino acids and pyrophosphate keep chromium in a diffusible form (Stokinger 1981).

3. Antagonisms and Synergisms

The possibility of an antagonism between chromium and vanadium was reviewed by Nielsen et al. (1980).

4. Physiological Requirements

About 1 μg absorbable Cr(III) per day is required for maintaining normal glucose metabolism (NAS 1980). Chromium deficiency has been reported in humans as well as other animals. Glucose tolerance factor may be a cofactor for the initiation of peripheral insulin action (Langard and Norseth 1977).

D. Specific Organs and Systems

1. Nose, Throat, and Sinuses

Workers exposed to chromates and chromic acid mist may develop nasal membrane inflammation and ulceration, and nasal septum perforation. Chronic rhinitis, laryngitis, and pharyngitis are also common. Polyps and hoarseness of the larynx and polyps or cysts of the sinuses were not quite as common (Stokinger 1981).

Nasal ulceration occurs at workplace air concentrations of chromic acid slightly above the current Federal ceiling standard (0.1 ng/m^3) (NIOSH 1973).

2. Lung

Progressive pulmonary fibrosis in a small number of workers was reported in 1962, but no subsequent reports have appeared (Stokinger 1981). Pneumoconiosis (either nodular or nonnodular) has also been reported. Bronchial asthma is common among chromate workers. See also Part IV. G.

3. Skin

Slow-to-heal skin ulcers affect about half of the workers exposed to soluble chromates or chromic acid. Contact dermatitis and sensitization from chromic acid and chromates is less prevalent with present-day work practices than reported 20 to 40 yr ago (Stokinger 1981).

4. Gastrointestinal Tract

Afflictions of the gastrointestinal tract of workers include decreased or loss of senses of taste and smell, gastrointestinal ulcers, and hypertrophic gastritis (Stokinger 1981).

E. Teratogenicity

Mild achondroplasia resulted in chicks to whose eggs was added 2.5 mg Na dichromate on the 8th day of incubation. Giving hamsters intravenous doses of Cr_2O_3 on day 7, 8, or 9 of gestation caused cleft palate in the fetuses and resorption (Shepard 1980).

F. Mutagenicity

Sodium, potassium, and calcium chromates are mutagenic in bacteria. Calcium chromate induces transformations of cell cultures (Langard and Norseth 1977).

G. Carcinogenicity

Observed/expected ratios of 5-40:1 have been reported for incidences of lung cancer in workers exposed to Cr(VI) compounds. Lung cancer is highest among heavy cigarette smokers. Cancer is also sometimes reported at other sites (Langard and Norseth 1977).

V. REFERENCES

ACGIH. 1983. TLVs® threshold limit values for chemical substances and physical agents in the work environment with intended changes for 1983-84. Cincinnati, Ohio: American Conference of Governmental Industrial Hygienists.

Langard S, Norseth T. 1977. Chromium. In: Toxicology of metals - Vol. II. Springfield, VA: National Technical Information Service, pp. 164-187. PB-268 324.

Lauwerys RR. 1983. Chapter II. Biological monitoring of exposure to inor- ganic and organometallic substances. In: Industrial chemical exposure: Guidelines for biological monitoring. Davis CA: Biomedical Publications, pp. 9-50.

Merck Index. 1983. 10th ed. Rahway, New Jersey: Merck & Co., Inc.

NAS. 1974. National Academy of Sciences. Chromium. Washington, DC.

NAS. 1977. National Academy of Sciences. Drinking water and health.
Vol. I. Washington, DC.

NAS. 1980. National Academy of Sciences. Recommended dietary allowances,
9th ed. Washington, DC.

Nielsen FH, Hunt CD, Uthus EO. 1980. Interactions between essential trace
and ultratrace elements. Ann NY Acad Sci 355:152-164.

NIOSH. 1973. Natl. Inst. Occupational Safety and Health. Criteria for a
recommended standard: occupational exposure to chromic acid. Washington,
DC: U.S. Government Printing Office. HSM 373-11021.

NIOSH. 1975. Natl. Inst. Occupational Safety and Health. Criteria for a
recommended standard: occupational exposure to chromium(VI). Washington,
DC: U.S. Government Printing Office. HEW Pub. NIOSH 76-129.

OSHA. 1981. Occupational Safety and Health Admin. Occupational safety and
health standards. Subpart 2--Toxic and hazardous substances. Code of federa
regulations 29 (Part 1910.1000) pp. 673-679.

Shepard TH. 1980. Catalog of teratogenic agents, 3rd ed. Baltimore, MD:
The Johns Hopkins University Press.

Snyder WS, Cook MJ, Nasset ES, Karhausen LR, Howells GP, Tipton IH. 1975.
International Commission on Radiological Protection. Report of the task
group on reference man. New York. ICRP Publication 23.

Stokinger HE. 1981. Chapter 29. The metals. In: Patty's industrial
hygiene and toxicology. 3rd ed. Volume IIA. Clayton GD, Clayton FE, eds.
New York: A Wiley-Interscience Publication. John Wiley and Sons,
pp. 1493-2060.

USEPA. 1980. Ambient water qualty criteria for chromium. Springfield,
Virginia: National Technical Information Service. PB81-117467.

COBALT

MAMMALIAN TOXICITY SUMMARY

This brief review of cobalt production, uses, chemistry, and toxicity has been largely excerpted and condensed from the uncopyrighted book edited by Smith and Carson (1980), based on their 1979 Midwest Research Institute report to the National Institute of Environmental Health Sciences (Contract No. NO1-ES-2-2090).

I. INTRODUCTION

A. Production

Cobalt is not currently mined in the United States although imported cobalt-containing nickel laterite concentrates are processed in Louisiana.

B. Uses

Superalloys (alloys based on iron, nickel, or cobalt that maintain their strength at high temperatures) used in jet engines, magnet alloys, and the salts and paint driers are the major cobalt consumption categories. Each account for approximately 20 to 30% of total U.S. consumption.

Cemented carbide (hard metal) tools are produced primarily from tungsten carbide (WC) and 2 to 25% cobalt powders (the "cement"). Their major use is for metal-cutting operations. The second largest use is for mining and quarrying tools, a use expected to grow because of the urgent need for fossil fuels. The last year for which data were available on consumption of cobalt in cemented carbides (299 MT) was 1969.

An important use of cobalt metal is in hard-facing materials. Hard-facing is the application by welding of a specific alloy to a part, primarily to deter wear. About 180 to 450 MT cobalt has been used annually in hard-facing alloys since the Korean War. Cobalt-containing hard-facing alloys, however, comprise only 1 to 2% of the world market for hard-facing materials.

Since 1967, between 260 and 550 MT cobalt has been used annually in high-speed tool steels (5 to 12% Co), hot-work tool steel ($\leq$ 4.25% Co), cold-work tool steel ($\leq$ 3.25% Co), maraging steels, high-temperature stainless steels (0.2% Co), and miscellaneous alloys such as quench-hardened steels (4% Co). High-speed tool steels are used principally for cutting tools in high-speed machining.

Cobalt salts and soaps are the most active and generally useful paint driers. They are used not only in oil-based paints but also in the newer latex paints that have been oil- or alkyd-modified and in unmodified

butadiene-resin latex paint systems. Up to 0.25% cobalt is added to the vehicle. The soaps themselves are provided as formulations with 6 to 12% cobalt. Cobalt soaps are also used as catalysts to promote curing or hardening
of unsaturated polyester resins in building glass fiber laminates, foundry
core oil binders, composition board binders, and silicone resins. In the
period 1975 to 1978, U.S consumption of cobalt in driers fluctuated from 350
to 1,200 MT/year.

Cobalt compounds are also used commercially as catalysts in the
oxo synthesis (hydroformylation of olefins), petroleum refining, and oxidation of organic compounds. Cobalt carbonyls--$Co_2(CO)_8$ and $HCo(CO)_4$--or a
cobalt-phosphine complex are the oxo synthesis catalysts. Major products
produced in the presence of cobalt catalysts include terephthalic acid (by
oxidation of p-xylene), which is used in the production of polyester fibers,
and phenol (via benzoic acid produced by the cobalt-catalyzed oxidation of
toluene [2% of U.S. phenol capacity in 1975] or via cumene peroxidation [88%
of U.S. phenol capacity]).

Cobalt compounds have important roles as ground coat frits, de-
colorizers, and pigments. Up to 3% cobalt (usually 0.5 to 0.6%) as a salt or
oxide is used in the blue-black ground coat or frit necessary for porcelain
enameling of steel for bathroom fixtures, large appliances, and kitchenware.
Addition of < 0.02% blue cobalt compounds to pottery, glazes, and enamels
masks the yellowish tinge imparted by impurities such as iron oxide. Several
oxide, silicate, and aluminate combinations of cobalt with other metal oxides
are used as pigments for porcelain decoration. Other cobalt pigments are
used in artists' oil paints and in printing inks for fabric and paper. Bank
notes are printed with cobalt-pigmented inks. Cobalt blue (cobalt silicate
or aluminate) is also used as a pigment for hydraulic cement.

C. Physical and Chemical Properties

Cobalt, atomic number 27, belongs to the group of elements compris-
ing the triad of iron, cobalt, and nickel. Cobalt is a blue-white hard metal
that melts at 1493°C, boils at 3100°C, and has a density of 8.90 g/cm^3. The
uncomplexed cobaltous ion (Co^{2+}) is stable in aqueous solution, but the co-
baltic ion (Co^{3+}) is a powerful oxidizing agent unless it is present in an
anhydrous crystal lattice or is complexed with various ligands. Cobaltous
oxide is probably the ultimate thermal decomposition product of any cobalt
species in air prior to its own decomposition at 1800°C to the elements. The
cobalt-cobaltic oxide, Co_3O_4, is the predominant product below 300°C in
oxygen or air (Smith and Carson 1980).

II. EXPOSURE AND EXPOSURE LIMITS

A. Oral

The concentration of cobalt in foods leads to an estimate of about
40 to 50 μg as the bulk of the daily oral intake (Smith and Carson 1980).
Mollusks and fish may bioconcentrate cobalt from contaminated waters by a
factor of several thousand, but bioconcentration in the terrestrial food

chain is low. Dietary standards recommend a daily intake of 5 µg vitamin B_{12} (containing 0.2 µg Co) (Herndon et al. 1980).

Drinking water concentrations average 2 µg/L although concentrations as high as 107 µg/L have been reported. Coffee contains 5 to 10 µg/cup. Beer contains 2 to 50 µg/L (natural content).

Chronic drinking water studies with rats at 2 mg Co/L have been observed to have an erythropoietic effect, to cause immunosuppression, and to inhibit reflex learning. Concentrations of 200 µg/L exhibited no such effects. In the USSR, a limit of 1 mg Co/L in drinking water has been suggested.

Anemic patients have been treated with cobalt salts at oral dose levels of 0.17 to 3.9 mg Co/kg. The smallest dose represents a daily intake of 12 mg/70 kg adult. The higher doses were associated with toxic symptoms including nerve deafness and goiter.

B. Inhalation

Probable major sources of cobalt in the atmosphere are coal and residual fuel oil burning and attrition in use of superalloys, hard-facing alloys, and cemented tungsten carbides. The chief form of cobalt entering the environment from atmospheric emissions is probably CoO.

Levels as high as 610 ng Co/m³ were found in ambient air in Cleveland, Ohio, a city with many metallurgical works' but average concentrations of cobalt in the major cobalt-refining town of Clydach, Wales, were only 48 ng/m³ (Smith and Carson 1980).

The permissible exposure limit in the United States for cobalt metal, dust, and fume as adopted by the Occupational Safety and Health Administration (OSHA) is 0.1 mg/m³, which is a time-weighted average (TWA) for a conventional workday (OSHA 1981). Several countries have set the same limit; ACGIH (1983) recommends the same limit as a TLV, but the USSR, Bulgaria, Poland, and the Federal Republic of Germany have set a standard of 0.5 mg Co/m³. In the USSR, MAC's have also been set for cobalt oxide (0.5 mg/m³), dicobalt octacarbonyl (0.01 mg/m³), and cobalt hydrocarbonyl and its decomposition products (0.01 mg/m³) (Herndon et al. 1980). In Belgium, the occupational limit for cobalt and cobalt oxide in air is 0.01 mg/m³ (NIOSH 1982).

C. Dermal

Cobalt hypersensitivity develops in a small percentage of exposed persons. Human dermal contact with cobalt may occur by use or work with alloys, glass pigment, ceramics, paint, rubber, and cobalt-containing drugs. In addition, cobalt is a common contaminant in the widely used metal, nickel, which is also a skin sensitizer. Farm workers are exposed to the cobalt-containing salt licks and feeds for ruminants and soil supplements. Cement workers also show cobalt sensitivity. Magnetic tape and polyester (produced by a cobalt-catalyzed reaction) are common in consumer items that could lead to skin contact and dermatitis.

D. Total Body Burden and Balance Information

Smith and Carson (1980) estimated that the total body burden of a 70-kg human is 1.5 mg cobalt (0.02 mg/kg). Snyder et al. (1975) estimated < 1.5 mg. Lung tissue appears to be a good indicator of above-normal exposure to cobalt in the air (Smith and Carson 1980).

Snyder et al. (1975) gave the following balance information based on literature values:

	µg Co/Day
Intake	
Food and fluids	300
Airborne	< 0.3
Excretion	
Urine	200
Feces	90
Sweat	4.0
Hair	2.4

Smith and Carson (1980) chose 20 to 40 µg Co/day as a better representation of the dietary intake.

III. TOXICOKINETICS

A. Absorption

1. Oral

Human absorption of cobalt from the diet is variable, appearing to average about 80% in some studies, but being as low as 1 to 44% in other studies (especially when given as an inorganic salt).

Vitamin B_{12} absorption depends on the presence of hydrochloric acid and the intrinsic factor, a mucopolysaccharide secreted by the parietal cells of the gastric mucosa. Without the intrinsic factor, B_{12} absorption is < 2% compared to approximately 70% in its presence (Herndon et al. 1980).

2. Skin

A scientist contaminated by the cobalt radioisotope [58]Co showed cobalt absorption by the intact skin.

Tests with guinea pigs showed that cobalt intake by skin absorption is low: only 1.72% of a dose of radioactive cobalt applied to the intact skin appeared in the urine and 0.9% appeared in the liver (Herndon et al. 1980

3. Lungs

A volunteer exposed to a gaseous mixture of several radioisotopes absorbed ^{60}Co via the lungs (Herndon et al. 1980).

B. Distribution

1. General

Cobalt occurs at a high level in the liver, presumably because of its portal circulation. Muscle, spleen, kidney, thymus, adrenals, pancreas, lungs, stomach wall, and bone have each also been reported to be the organ with the highest concentration or total content of cobalt in various distribution studies. Generally, cobalt given either parenterally or orally does not appear to accumulate in a specific target organ in many species.

From their extensive literature compilation of cobalt in human tissues, Cole and Carson (1980) gave the following values as representative: lungs, 0.2; hair, 0.5; kidneys, 0.1; liver, 1; brain, 0.1; heart, 0.2; intestine, 0.8; muscle, 1; spleen, 1; skin, 0.03 mg/kg dry weight. About 50 to 90% of the total body store of vitamin B_{12} is in the liver.

2. Blood

The concentrations of cobalt carried in the blood appear to be highly variable. Perhaps some of the analytical methods used have distinct limitations for cobalt detection in this matrix or perhaps exposure to cobalt from sources other than diet and drinking water also are highly variable.* Cole and Carson (1980) compiled the followng ranges for reported mean values for cobalt in whole blood and blood fractions of unexposed individuals.

Whole blood	0.5-238 µg/L
Plasma	Not detected - 43 µg/L
Serum	0.2-660 µg/L
Red blood cells	10.5-206 µg/L

Lauwerys (1983) gives 100 to 750 µg/L as normal urine values.

3. Adipose

Forbes et al. (1954; cited by Cole and Carson 1980) determined cobalt in the tissues of a 46-year-old U.S. male. The cobalt concentration in his adipose tissue was 0.008 mg/kg fresh weight.

Cole and Carson (1980) picked 0.1 mg/kg dry weight as a typical value for adipose tissue.

---

* Cobalt values in blood serum reported in the literature are often grossly in error due to sample contamination according to Versieck (1984).

Snyder et al. (1975) give the following values for the total amount of cobalt in the adipose tissue of Reference Man:

	μg	μg/g
Subcutaneous	250	0.017
Other separable	< 30	0.002
	(< 9.0 to 150)	(< 0.0006 to 0.010)
Interstitial	32	0.0021

These values do not agree well with the 48 μg cobalt in adipose tissues reported by Forbes et al. (1954; cited by Cole and Carson 1980).

C. Excretion

Cobalt is excreted in both urine and feces of humans. The cobalt that is absorbed is largely excreted by the urine with some (perhaps ~ 5% of that absorbed) excretion via the bile and feces. The large fraction of ingested cobalt eliminated via the feces indicates that much of it is probably unabsorbed. Cobalt balance studies in both humans and animals show great variability. Whether this reflects biological variability in cobalt absorption and excretion or analytical problems is not clear.

Once cobalt reaches the tissues, elimination is slow. Some body pools of radiocobalt have been reported to have biological half-lives ranging up to 69 years. The longest biological half-lives have been observed after inhalation exposure. The very long half-lives may be due to uptake by pulmonary lymph nodes. A study of long-term excretion of cobalt in rats showed that the bone ultimately contained the largest stores of the intravenously dosed cobalt.

Perdrix et al. (1983) and Pellet et al. (1984) found that 24-h urinary cobalt concentrations correlated well with exposure to cemented tungsten carbide containing up to 20% Co.

The main route of vitamin B_{12} excretion is by way of the bile. The biological half-life of vitamin B_{12} is about 400 days (Herndon et al. 1980).

IV. EFFECTS

A. Acute and Other Short-Term Exposures

Patients receiving 0.17 to 3.9 mg Co/kg/day for 6 days to 8 mo, usually for the treatment of anemia, showed 20 to > 90% depression in iodine uptake, which was typical of disease states producing an inhibition of the protein binding of inorganic iodide. Doses as high as 200 mg $CoCl_2$/day were given. Goiters and classic signs of hypothyroidism occurred as early as 6 wk after initiation of cobalt therapy. By 1970, all antianemia preparations containing cobalt had been removed from the U.S. market. Other symptoms that were seen in patients receiving cobalt antianemia preparations included anorexia, nausea, vomiting, diarrhea, substernal aches, erythema and hot sensations, skin rashes, and tinnitus and neurogenic deafness.

Other manifestations of cobalt neurotoxicity include peripheral neuritis, headaches, weakness and irritability, changes in the electrical activity of the brain, and changes in reflexes, especially visual. (These transient symptoms were seen in 118 workers suffering from acute industrial exposure to cobalt carbonyl vapor.) Optic nerve atrophy has been described in at least two patients dosed with total of 0.9 and 32 g $CoCl_2$ (Herndon et al. 1980).

Lethal doses of cobalt metal and compounds in laboratory mammals are summarized in Table 2 [references cited are from Smith and Carson 1980]. The oral LD_{50}'s or LD_{LO}'s (lowest lethal doses) in rats for cobalt metal, the carbonate, cobaltous oxide, $Na_2[Co-EDTA]$, and the stearate ranged from 1,000 to 2,000 mg/kg. Co_2O_3 is even less toxic, having an intraperitoneal LD_{50} of 5,000 mg/kg.

Cobalt compounds having LD_{50}'s or LD_{LO}'s in the range 100 to 1,000 mg/kg when given orally to rats include the tetracarbonyl, the nitrate, the chloride, the fluoborate, and the sulfate. The lactate is more toxic when given intraperitoneally (7.5 mg/kg) than is the chloride given intravenously (20 mg/kg).

Mammals acutely poisoned by cobalt salts exhibit cutaneous vaso-dilation, diarrhea, loss of appetite, paralysis of hindlegs, and albuminuria or anuria. Histopathological investigation reveals congestion of all organs with large hemorrhages in the liver and adrenals, hyperplastic bone marrow, alveolar thickening of the lungs, tubular degeneration in the kidneys, pancreatic degenerative changes, and pale and shrunken myocardial fibers (Stokinger 1981).

B. Chronic Exposure

Cobalt exhibits toxic effects in humans on the thyroid (see Section IV.A), the heart, and, possibly the kidney, in addition to allergic manifestations and the occupational lung disease seen in the cemented carbide industry (Herndon et al. 1980).

The epidemics* of severe congestive heart failure with > 40% mortal-ity in heavy beer drinkers from addition of approximately 1 mg Co/L to beer as a foam stabilizer were apparently associated in a causal way with chronic alcoholism (Herndon et al. 1980).

Common complaints among workers occupationally exposed to cobalt powders include partial or complete loss of the sense of smell (anosmia), dyspnea, gastrointestinal distress, and weight loss. Lung fibrotic changes are attributed to the action of cobalt even when cobalt is associated with tungsten and titanium carbides (Herndon et al. 1980).

---

* Documented in Louvain, Belgium; Omaha, Nebraska; Minneapolis, Minnesota; and Quebec City, Canada.

TABLE 2

ACUTE TOXICITY TO LABORATORY MAMMALS (LETHAL DOSE LEVELS)

Species, Strain	Cobalt Compound	Dose	Route	Comments	Reference
Piebald rat	Co (metal)	25 mg/kg	I.P.	100% survival with respiratory symptoms	Harding (1950)
Rat, white	Co, powdered metal	100 to 200 mg/kg	I.P.	LD_{50}; dose is age dependent	Frederick and Bradley (1946)
Rabbit	Co (metal)	100 mg/kg	I.V.	Lowest lethal dose for species and route	Fairchild et al. (1977)
Rat	Co (metal)	100 mg/kg	I.V.	Lowest lethal dose for species and route	Fairchild et al. (1977)
Rat	Co (metal)	25 mg/kg	Intratracheal	Lowest lethal dose for species and route	Fairchild et al. (1977)
Piebald rat	Co tetracarbonyl	12.5 mg/kg	Intratracheal	Lethargy and death to all in 15 min to 6 hr	Harding (1950)
Rat	Co tetracarbonyl	753.8 mg/kg	Intragastric	LD_{50} dose range: 675 to 832.6 mg/kg	Spiridonova and Shabalina (1973)
Mouse	Co tetracarbonyl	377.7 mg/kg	Intragastric	LD_{50} dose range: 281.7 to 473.7 mg/kg; hemodynamic and dystrophic changes of internal organs; local damage to gastric mucosa	Spiridonova and Shabalina (1973)
Rat, Carworth-Wistar	CoO	1,700 mg/kg	Oral	LD_{50}: 1,070 to 2,828 mg/kg = 95% confidence limits	Smyth et al. (1969)
Mouse, white	CoO	19.3 mg/kg	"Injection"	"Threshold dose" based on six organ effects	Rabolnikova (1971)
Rat, white	Co_2O_3	5,000 mg/kg	I.P.	LD_{50}	Frederick and Bradley (1946)
Mouse, white	Co_2O_3	242 mg/kg	"Injection"	"Threshold dose"	Rabolnikova (1971)
Rat	$CoCO_3$	2,002 mg/kg	I.P.	Lowest lethal dose for species and route	Fairchild et al. (1977)
Rat, male	$CoSO_4$	597 mg Co/kg	Oral	LD_{50}	Wiberg et al. (1969)
Rat	$CoSO_4$	13 mg Co/kg	I.P.	LD_{50}	Wiberg et al. (1969)
Mouse, white Swiss male	$CoSO_4$	21 mg Co/kg	I.P.	LD_{50} (30-day)	Bienvenu et al. (1963)
Dog	$CoSO_4 \cdot 7H_2O$	16 mg/kg	I.V.	Lowest lethal dose for species and route	Fairchild et al. (1977)
Rat, spontaneously hypertensive	"Common soluble salt"	13.8 mg/kg	I.P.	7-day LD_{50}	Lewis (1975)
Rabbit	$Co(NO_3)_2$	400 mg/kg	Oral	Lowest lethal dose for species and route	Fairchild et al. (1977)
Rat	Cobalt stearate	2,390 mg/kg	Oral	LD_{50}	Schmidt et al. (1975)
Rat	Tris(monoethanolamine) cobalt	408.0 mg/kg	S.C.	LD_{50}	Chernov et al. (1973)
Rat, young	$CoCl_2 \cdot 6H_2O$	50.0 mg Co/kg of diet	Oral	25% mortality (corn starch, skim milk diet)	Huck (1975)
Rat, young	$CoCl_2 \cdot 6H_2O$	400.0 mg Co/kg of diet	Oral	100% mortality (corn starch, skim milk diet)	Huck (1975)
Rat	$CoCl_2$	80 mg Co/kg	Oral	LD_{50}	Krasovskii and Fridlyand (1971)
Rat	$CoCl_2$	144 mg Co/kg	Oral	LD_{50}	Murdock and Klotz (1959)
Rat, Wistar	$CoCl_2$	860 mg/kg	Oral	24 hr LD_{50} dose range: 750 to 980 mg/kg	Puget et al. (1975)
Mouse, Swiss	$CoCl_2$	620 mg/kg	Oral	24 hr LD_{50} dose range: 570 to 690 mg/kg	Puget et al. (1975)
Mouse	$CoCl_2$	80 mg Co/kg	Oral	LD_{50}	Krasovskii and Fridlyand (1971)

TABLE 2 (Concluded)

Species, Strain	Cobalt Compound	Dose	Route	Comments	Reference
Rat, Wistar	$CoCl_2$	140 mg/kg	S.C.	24 hr LD_{50} dose range: 110 to 170 mg/kg	Puget et al. (1975)
Mouse, Swiss	$CoCl_2$	160 mg/kg	S.C.	24 hr LD_{50} dose range: 140 to 180 mg/kg	Puget et al. (1975)
Mouse, female	$CoCl_2$	40.8 mg/kg	S.C.	LD_{50}	Hasegawa (1974)
Guinea pig	$CoCl_2$	2.0 mL of 0.239 M solution (30 mg)	P.C.	~ LD_{50}: 11/20 dead in 3 weeks	Wahlberg (1965)
Rat, Wistar	$CoCl_2$	50 mg/kg	I.P.	24 hr LD_{50} dose range: 40 to 60 mg/kg	Puget et al. (1975)
Rat	$CoCl_2$	7.9 mg Co/kg	I.P.	LD_{50}	Murdock and Klotz (1959)
Mouse, male	$CoCl_2$	36.7 mg/kg	I.P.	LD_{50}	Hasegawa (1974)
Mouse, Swiss	$CoCl_2$	130 mg/kg	I.P.	24 hr LD_{50} dose range: 110 to 150 mg/kg	Puget et al. (1975)
Guinea pig	$CoCl_2$	2.0 mL of 0.239 M solution (30 mg)	I.P.	~ LD_{50}: 6/11 dead in 24 hr	Wahlberg (1965)
Rat	$CoCl_2$	20 mg/kg	I.V.	LD_{50}	Fairchild et al. (1977)
Rat, white	Co lactate	7.5 mg/kg	I.P.	LD_{50}	Frederick and Bradley (1946)
Rat	Co niacinamide	134 mg/kg	Oral	LD_{50}	Murdock and Klotz (1959)
Dog	$[Co(NH_3)_6]$ nicotinate	45 mg/kg range	I.V.	LD_{50}; hypotension and tachycardia noted with this trivalent Co salt	Polonovski et al. (1954)
Rat	$Na_2[Co-EDTA]$	> 1,000 mg Co/kg	Oral		Murdock and Klotz (1959)
Rat	$Na_2[Co-EDTA]$	> 800 mg Co/kg	I.P.		Murdock and Klotz (1959)

Long-term exposures in the cemented tungsten carbide industry give rise to extrinsic asthma, diffuse interstitial pneumonitis with a 50% reduction in vital capacity, fibrosis, and/or pneumoconiosis. About 5% of the respiratory complaints are allergic in nature and show improvement away from the work environment. Of $\sim$ 5,000 workers exposed to cemented tungsten carbides described in the literature, $\sim$ 500 showed various symptoms of respiratory toxicity. Cobalt concentrations in the workplace air have often exceeded the current OSHA permissible exposure limit of 0.1 mg/m^3. Concentrations as high as 79 mg Co/m^3 have been reported in the older literature (Herndon et al. 1980).

Although cobalt metal presumably represents the bulk of occupational cobalt inhalation exposures, the oxides and salts may also produce intoxication. A hydrometallurgical cobalt plant may produce aerosols containing, for example, CoNiS, $Co(NH_3)_6CO_3$, $[Co(NH_3)_5SO]_2SO_3 \cdot 2H_2O$, $Co(OH)_2$, $CoSO_4$, and Co metal. At one such plant, workers absorbed at least 694 µg/day since this was the average urinary excretion. Air concentrations at this plant ranged from 0.063 to 1.605 mg Co/m^3. Blood values of erythropoietin, red blood cell count, etc., were not significantly different from those of unexposed controls, but the albumin and lactic dehydrogenase isoenzyme were highly significantly greater in the exposed group. Lung disease was not reported (Herndon et al. 1980).

Among > 4,200 workers surveyed in six studies of the cemented tungsten carbide industry, 1.7 to 9.4% of the populations showed dermal hypersensitivity or respiratory allergic symptoms. An erythematous, papular type of dermatitis, most marked in friction areas, and positive patch tests to metallic cobalt powder were often reported. Besides occupational exposures in the manufacture of cobalt-containing alloys, coatings, pigmented glass and ceramics, paint, and drugs, farm workers may show an allergic dermatitis from handling the cobalt nutrient preparations used in agriculture (Herndon et al. 1980).

C. Biochemistry

1. Effects on Enzymes

Although the mode of action of cobalt is not fully understood, a large part of its effects are produced by enzymatic impairment which leads to depressed tissue respiration and further depressive effects on energy metabolism. For example, Co^{2+} blocks the Krebs citric acid cycle and cellular respiration. It also depresses some drug-metabolizing enzymes of the liver and stimulates others.

The cobaltous ion can substitute for other divalent cations in several enzymes. For example, Co^{2+} can substitute for Zn^{2+} and activate alcohol dehydrogenase, lactate dehydrogenase, carboxypeptidase A, carbonic anhydrase, and alkaline phosphatase.

Besides reducing drug-oxidizing activity, cobalt affects other liver enzymes. Cobalt is the most efficient ion for inducing heme oxygenase in the liver. It also affects hepatic catalase and the mitochondrial enzyme aminolevulinic acid synthetase. Animals dosed with cobalt show elevated serum concentrations of lactate dehydrogenase, serum glutamic oxaloacetic transaminase (SGOT), and serum glutamic pyruvic transaminase (SGPT). The lipemic response (elevated plasma triglycerides, cholesterol, and lipoproteins seen in animals treated with ~ 10 mg $CoCl_2$/kg for 11 to 12 days) may be due to cobalt inhibition of the enzyme responsible for catabolism of circulating lipoproteins.

Vitamin B_{12} is a coenzyme for three enzyme systems in animals and bacteria. These systems play important roles in metabolizing propionate, methylating homocysteine to give methionine, and reducing ribonucleotides. Ruminal microflora of multigastric animals convert dietary cobalt to vitamin B_{12}, but man and other monogastric species depend on ruminants and bacteria for the production of vitamin B_{12} (Herndon et al. 1980).

2. Metabolism

Cobalt reacts with the thiol groups of amino acids, proteins, and other coenzymes or cofactors, thereby inactivating several biochemical pathways, especially those associated with cellular energy production. Vitamin B_{12} (cyanocobalamin) metabolism is intimately combined with that of folic acid. By methyl transfers from 5-methyltetrahydrofolate to vitamin B_{12} coenzyme and from the methylcobalamin formed to homocysteine, tetrahydrofolic acid is recycled back to the folate pool and the amino acid methionine is synthesized (Herndon et al. 1980).

3. Antagonisms and Synergisms

Administration of cobalt salts to animals and humans increases the formation of red blood cells. This may not be a normal physiologic stimulus of erythrocyte synthesis but rather may be a toxic manifestation of cobalt activity. When cobalt inhibits tissue respiration and oxidative phosphorylation, tissue hypoxia occurs. This hypoxia is the probable stimulus to the secretion of erythropoietin, the hormone that stimulates bone marrow stem cells to produce erythrocytes.

Cobalt appears to have a synergistic toxic effect with ethanol on the heart and on general growth rate.

The lung toxicities of cobalt and other ingredients of cemented tungsten carbide are greater together than when administered separately.

Selenite and vitamin E protect against the myocardial toxicity of cobalt. Cobalt protects against the testicular necrosis caused by cadmium. Animals ingesting ≥ 4 mg Co/kg show an anemia indicative of depression of iron absorption.

Cobalt deficiencies render sheep more susceptible to selenium toxicity and children more sensitive to the toxicity of vitamin D (Herndon et al. 1980).

4. Physiologic Requirements

Vitamin B_{12} concentrations of 0.2 µg/L are sufficient to prevent any signs of clinical, hematological, or biochemical deficiencies in humans. Malabsorption is the primary cause of the B_{12} deficiency anemias rather than deficient dietary intakes except when animal protein intake is low.

Deficiency of vitamin B_{12} or folate results in defective synthesis of DNA, which is manifested early in hemopoietic tissue where megaloblastic changes are seen in developing red and white cells and in the megakaryocytes. Vitamin B_{12} deficiency also causes demyelinating neurological lesions of unknown biochemical etiology.

The minimum daily dose of cobalt for most animals is 1 to 10 µg except for ruminants whose daily requirement is 100 µg. Cobalt deficiency diseases in ruminants have been observed wherever the herbage is very low in cobalt due to inadequate soil concentrations. Ruminants require cobalt so that their ruminal microflora can synthesize the vitamin B_{12} that is required for propionate metabolism, etc. Propionate metabolism is more important to energy metabolism in ruminants than in nonruminants. Deficiency symptoms in ruminants include wasting with anemia and anorexia.

Cobalt compounds (acetate, carbonate, chloride, oxide, and sulfate) are commonly added to animal feeds, to the salt lick, or to fertilizers. Cobalt pellets placed into the rumen are also effective in preventing cobalt deficiency in ruminants.

Although cobalt is only known to be required by nonruminants as a constituent of vitamin B_{12}, some evidence points to a cobalt requirement for the optimal utilization of low doses of iodine by the thyroid (Herndon et al. 1980).

D. Specific Organs and Tissues

1. Heart

Interference of cobalt with cardiac metabolism produces dilatation of the heart and secondary thrombosis. The appearance of the myocardial toxicity observed in a 41-year-old hard metal worker exposed to cobalt for 6 yr resembled those of the fatal cases of beer-drinkers' cardiomyopathy: fragmented myocardial fibers, vacuolar change, diffused thickening of the endocardium, and absence of an inflammatory reaction. Five cases of myocarditis were also reported in cobalt smelter workers (Herndon et al. 1980).

Numerous animal studies on the role of cobalt in cardiomyopathy reviewed by Herndon et al. (1980) indicate that the mechanism of cobalt toxicity to the rat heart is based on a cobalt-induced reduction in the oxidation of pyruvate. Animal pathology is more pronounced if thiamine and protein deficiency is present.

2. Hematopoietic System

Cobalt treatment of anemic humans at doses up to 200 mg $CoCl_2$/day for up to 44 wk significantly increased the erythrocyte count, attaining maximum values within 2 to 3 mo and returning to pretreatment levels within 12 wk after cessation of cobalt therapy.

Several animal studies indicate that the erythropoietic effect of cobalt is a toxic rather than beneficial effect.

An erythropoietic effect was seen after 6 to 7 mo in rats drinking water containing $\geq \sim$ 2 mg Co/L as mentioned above. Rats exposed to cobalt metal aerosols of 0.05 mg/m^3 showed initially depressed but ultimately elevated hemoglobin and erythrocyte count after 2.5 mo. Rats exposed to 0.005 mg Co/m^3 showed an increase only in hemoglobin, but pathomorphological changes were seen in lungs and other organs even at this level. Toxic effects included decreased activity of lymphatic tissue in spleen, decreased reactivity of bronchial cell membranes and lymphatic tissue in lungs, and protein dystrophy in kidneys. Protein and carbohydrate metabolisms were also disturbed.

The mechanisms of the erythropoietic effect appear to involve early decomposition of both old and newly formed erythrocytes under the action of hypoxia at higher levels of cobalt exposure, which elicits the hemopoietic effect on bone marrow. Ten times the human dose of $CoCl_2$ injected into rats produced hyperplasia of the bone marrow, particularly in the erythropoietic elements (Herndon et al. 1980).

Alterations of blood clot formation may occur at levels found in the workplace (NIOSH 1982).

3. Thyroid

Thyroid function tests including ^{131}I uptake were depressed from 20% to > 90% in various studies of patients receiving 0.17 to 3.9 mg Co/kg/day for a duration of 6 days to 8 mo (Villaume et al. 1976; cited by Herndon et al. 1980). The uptake patterns were found to be typical of disease states producing an inhibition of the protein binding of inorganic iodide. In some case histories, goiters and classic signs of hypothyroidism occurred as early as 6 wk after initiation of cobalt treatment (usually for anemia). The thyroid glands of infants and children appear to have a special susceptibility to cobalt, but it is unrelated to the dose received. In tests with rats, either a cobalt deficiency or an excessive intake of cobalt produced a decrease in iodine in the thyroid gland (Herndon et al. 1980).

NIOSH (1982) recommended thyroid palpation as part of the medical surveillance of cobalt-exposed workers.

4. Lungs

According to the 1977 NIOSH criteria document, the most commonly reported symptoms after occupational exposure to cobalt in the cemented tungsten carbide industry are: upper respiratory tract irritation, exertional

dyspnea, coughing, weight loss, extrinsic asthma, diffuse interstitial fibrosis, and pneumoconiosis, alone or in combination. About 5% of the complaints are allergic in nature (Herndon et al. 1980). Workers exposed to
0.1 - 0.2 mg Co/m^3 show mild to fibrotic lung changes; those exposed to
0.06 mg/m^3 develop airway obstruction (NIOSH 1982).

Pigs breathing 0.1 mg Co metal/m^3 for 3 mo showed a lung compliance only 66% that of the controls, but compliance in the exposed group
returned to control levels within 1 mo after cessation of cobalt exposure.
Blood chemistry studies of the exposed pigs revealed an increase in α-, β-,
and γ-globulins over those of controls, a net increase in total protein, and
inversion of the albumin/globulin ratio. Such changes may be interpreted as
early indicators of lung cell damage. Pigs exposed to 0.1 mg Co/m^3 and 1.0
mg Co/m^3 excreted 29 and 220 µg Co/L, respectively, compared to 18 µg/L by
the controls. This study, showing serious effects with relatively low exposure, is cited in the 1977 printing of the 1971 edition of Documentation
of Threshold Limit Values for Workroom Air as the reason for suggesting a
lowering of the adopted TLV for cobalt metal dust and fume from 0.1 to 0.05
mg/m^3 (Herndon et al. 1980).

NIOSH (1982) considered it prudent to consider all cobalt compounds
as being capable of causing lung fibrosis.

5. Skin

The 1977 NIOSH criteria document on cemented tungsten carbide reported allergic dermal symptoms in 1.7 to 9.4% of the worker populations
studied. Dermal symptoms included an erythematous, papular type of dermatitis. Positive patch tests to metallic cobalt powder were often reported.
Skin eruptions were most marked in friction areas.

The dermal sensitivity to cobalt must be prevalent in the general
population, too. For example, in patch tests with 281 European domestic
workers, 8% gave a positive reaction to 2% $CoCl_2$ solution. Wahlberg (1973;
cited in Herndon et al. 1980), using serial dilutions of cobalt chloride,
determined that the mean threshold for 60 cobalt-allergic patients was 0.27%
in distilled water and 0.31% in petrolatum. He also noted, however, that 14
of the subjects were exquisitely sensitive to the cobalt challenge and at no
dilution was the threshold of sensitivity obtained.

An allergic etiology for the rejection of metal alloy implants (60
to 65% Co, 27% Cr, and 2.5% Ni) has been suggested. When 16 of 35 patients
who exhibited a severe reaction to their implants gave positive skin tests to
metals, 13 of the 16 were sensitive to cobalt.

Allergic sensitivity to vitamin B_{12} has also been reported.

Intraperitoneal or intravenous injections of 5 mg $CoSO_4$ produced in
rats erythema and edema of the ears, paws, and snout attributed to histamine
release by the mast cells in these tissues (Herndon et al. 1980).

6. Nervous System and Behavior

There are only brief references to nervous system toxicity in humans resulting from cobalt. A few references to ill-defined paresthesias, numbness, and other peripheral signs that accompanied more pronounced symptoms of cobalt effects.

Optic nerve damage was observed in two patients who had been treated with $CoCl_2$ (total 73 g Co in one case within 2.5 year and 900 mg $CoCl_2$ within 30 days in the other case).

Acute cobalt carbonyl vapor intoxication of 118 workers produced transient central nervous symptoms including headaches, weakness and irritability, changes in brain electrical activity, and changes in reflexes, especially visual ones (Herndon et al. 1980).

Brain implants of cobalt metal are commonly used to study epilepsy in animals, but the epileptogenic etiology of cobalt is unknown. In the rare cases of human epilepsy with a definite focus, the histopathology resembles cobalt epilepsy with abnormal neurons that have few dendrites.

Behavioral effects have been reported in rats drinking water containing $CoCl_2$ for 6 days/wk for 7 mo at doses of 0.5 mg/kg. This dose corresponds to a level of ~ 2,000 µg Co/L in the water. The effect on reflex learning (lengthening of the latent period and increase in the number of extinctions of conditioned reflexes) seen in rats drinking ~ 2000 µg Co/L was not seen in rats receiving water containing ~ 200 µg Co/L. These rats showed no differences from the behavior of controls (Herndon et al. 1980).

7. Pancreas

Intravenous injections of cobalt chloride in rats, dogs, rabbits, and guinea pigs produced severe toxicity (degranulation) or loss of the alpha cells of the pancreatic islets of Langerhans. The effect was accompanied by a transient hyperglycemia. For example, 1 to 4 h after rabbits were dosed with $CoCl_2$ at 35 mg/kg, the glycogen had disappeared almost entirely from the liver. The initial hyperglycemia may be due to some extrapancreatic factor whereas a subsequent increase in blood glucose may be ascribed to the degranulation of the beta cells induced by cobalt (Herndon et al. 1980).

8. Kidney

Cobaltous chloride treatment of the anemia of patients with chronic renal failure was recommended as recently as 1976. Other researchers, however, found that the whole blood cobalt levels were significantly higher in dialysis patients who had been treated with $CoCl_2$ 13 to 20 mo previously than in those who had not received cobalt. Since this prolonged retention of cobalt could lead to toxicity, these researchers discouraged the use of $CoCl_2$ (Herndon et al. 1980).

E. Teratogenicity

        Cobalt has not been shown to cause significant teratogenic or re-
productive sequelae in humans. Brill et al. (1974; cited by Herndon et al.
1980) found no differences between normal and congenitally abnormal fetuses
and infants and their mothers in metal levels, including cobalt, in blood
and/or tissues. Twenty pregnant women delivered normal children after receiving 75 to 100 mg $CoCl_2$/day for treatment of anemia. A group of pregnant
women who received cobalt-iron mixture (dose or number of subjects not reported) also all delivered normal children. No toxicity was reported in any
women receiving the cobalt either alone or in combination (Jacobziner and
Raybin 1961; cited by Herndon et al. 1980). There have been no reports of
human teratogenic or reproductive effects of cemented tungsten carbide dust
(which contains up to about 25% cobalt), although there is a large workplace
population at risk (NIOSH 1977).

        Cobaltous chloride and nitrite salt solutions induced fetal cleft
palates when injected alone into mouse dams but inhibited cleft formation
caused by cortisone injections. Birth defects in swine have been attributed
to dietary cobalt deficiencies. Decreased fertility is also associated with
cobalt deficiency.

        In vitro, 10 mg Co/L completely inhibited growth in embryonic chick
tissue. A level of 0.75 mg/kg embryo modified development of the endocrine
pancreas. Another study exposing chick embryos to $CoCl_2$ reported delayed organization patterns, over-development of neural and mesodermal structures,
and neural tube and notochord position abnormalities (Herndon et al. 1980).

F. Mutagenicity

        In cultures of human leukocytes and diploid fibroblasts, 40 μg Co/L
(a not unusual human serum concentration) as $Co(NO_3)_2$ markedly reduced the
mitotic index. Subtoxic doses added 2 to 24 h before fixation revealed no
chromosome aberrations. Human fibroblasts grown for several weeks in a
medium containing cobalt showed overcontracted chromosomes that were otherwise like those of the control cells. Cultured human lymphocytes exposed to
cobalt acetate at 0.06 to 0.6 μg/L showed increased frequency of diploidy
formation.

        No chromatid changes or chromosome aberrations were found in sur-
vivors of cattle acutely poisoned by cobalt when compared to the controls.
However, rats that developed tumors (rhabdomyosarcomas) after receiving
intramuscular injections of cobalt metal powder showed a great number of
abnormal cells and cells in mitosis. The cells specifically showed multi-
polarity, aberrant chromosomes, and failure of the chromosomes to separate in
anaphase and polyploidy.

        $CoCl_2$ tests with Bacillus subtilis strains H 17 Rec[+] and M 45 Rec[-]
indicated that cobalt was not DNA damaging. Cobalt was either strongly mutagenic or nonmutagenic, depending on cell density, in tests with Saccharomyces
cerevisiae (yeast). In higher plants (Allium cepa), cobalt treatment caused
sticky chromosomes (Herndon et al. 1980).

G. Carcinogenicity

Although current evidence is not conclusive, it appears that cobalt metal, cobalt salts, and cobalt carbonyl are not significant causative agents of human cancer. Industrial cases where malignancy has been associated with metals have usually involved exposure to a mixture, thereby preventing assignment of causation to one specific metal such as cobalt. Elinder and Friberg (1977; cited by Herndon et al. 1980) in their review of cobalt toxicity did not note any carcinogenic effects of cobalt metal or cobalt salts in humans.

Alveolar cell metaplasia and stages of lung fibrosis, however, have been observed in workers exposed to hard metal (tungsten or titanium carbide and cobalt cementing metal); and Bech et al. (1961; cited by NIOSH 1977) reported anaplastic adenocarcinoma of the right bronchial lobe in a 63-year-old male exposed to hard metal dusts for 17 yr (Herndon et al. 1980).

V. REFERENCES

ACGIH. 1983. TLVs® threshold limit values for chemical substances and physical agents in the work environment with intended changes in 1983-84. Cincinnati, Ohio: American Conference of Governmental Industrial Hygienists.

Cole CJ, Carson, BL. 1980. Chapter 7. Cobalt in the food chain. In: Trace metals in the environment. Volume 6 - Cobalt. Smith IC, Carson BL, eds. Ann Arbor, Michigan: Ann Arbor Science Publishers, Inc./The Butterworth Group, pp. 777-924.

Herndon BL, Jacob RA, McCann J. 1980. Chapter 8. Physiological effects. In: Trace metals in the environment. Volume 6 - Cobalt. Smith IC, Carson BL, eds. Ann Arbor, Michigan: Ann Arbor Science Publishers, Inc./The Butterworth Group, pp. 925-1140.

Lauwerys RR. 1983. Chapter II. Biological monitoring of exposure to inorganic and organometallic substances. In: Industrial chemical exposure: Guidelines for biological monitoring. Davis CA: Biomedical Publications, pp. 9-50.

NIOSH. 1977. Natl. Inst. Occupational Safety and Health. Criteria for a recommended standard: occupational exposure to tungsten and cemented tungsten carbide. Springfield, Virginia: U.S. Department of Commerce, National Technical Information Service. Pub. PB-275 594.

NIOSH. 1982. Natl. Inst. Occupational Safety and Health. Occupational hazard assessment: Criteria for controlling occupational exposure to cobalt. Washington, DC: U.S. Government Printing Office. DHHA (NIOSH) Publication No. 82-107.

OSHA. 1981. Occupational Safety and Health Admin. Occupational safety and health standards. Subpart 2--Toxic and hazardous substances. Code of federal regulations 29 (Part 1910.1000) pp. 673-679.

Pellet F, Perdrix A, Vincent M, Mallion JM. 1984. Biological determination of urinary cobalt. Significance in occupational medicine in the monitoring of exposures to sintered metallic carbides. Arch Mal Prof Med Trav Secur Soc 45(2):81-85; Chem Abstr 1984. 101:59461h.

Perdrix A, Pellet F, Vincent M, De Gaudemaris R, Mallion JM. 1983. Cobalt and sintered metal carbides. Value of the determination of cobalt as a tracer for exposure to hard metals. Toxicol Eur Res 5(5):233-240; Chem Abstr 1984. 100:179558w.

Smith IC, Carson BL, eds. 1980. Trace metals in the environment. Volume 6 - Cobalt. Ann Arbor, Michigan: Ann Arbor Science Publishers, Inc./The Butterworth Group, 1202 pp.

Snyder WS, Cook MJ, Nasset ES, Karhausen LR, Howells GP, Tipton IH. 1975. International Commission on Radiological Protection. Report of the task group on reference man. New York. ICRP Publication 23.

COPPER

MAMMALIAN TOXICITY SUMMARY

I. INTRODUCTION

A. Occurrence and Production

Native (elemental) copper was one of the first metals used by man. Bronze (copper hardened by alloying with tin) was the first "industrial metal;" its introduction ended the Stone Age, a major milestone in human history.

Copper is ubiquitous in the earth's crust, primarily found as sulfides and oxides. Major ores include chalcopyrite ($CuFeS_2$), cuprite (Cu_2O), malachite ($CuCO_3 \cdot Cu[OH]_2$), azurite ($2CuCO_3 \cdot Cu[OH]_2$), and bornite (Cu_5FeS_4). Metallic copper is prepared by smelting and electrolytic refining (Piscator 1977; Stokinger 1981).

B. Uses

About half of copper production is used as a conductor in electrical equipment due to its high conductivity. It is used in many alloys: beryllium-copper, brass, bronze, gunmetal, bell metal, german silver, etc. These are used in plumbing and the manufacture of various parts and goods. Copper compounds are used in pesticides, especially antifouling paints, algicides, fungicides, and insecticides. Some compounds are used as pigments in paints and ceramics (Piscator 1977; Stokinger 1981).

C. Chemical and Physical Properties

Copper forms two series of salts: cuprous (+1) and cupric (+2). Both oxidation states are involved in many stable complexes, such as $Cu(NH_3)_6Cl_2$ (Stokinger 1981). Salts may hydrolyze at ambient pH values and be precipitated as $CuCO_3$ (NAS 1977).

II. EXPOSURE AND EXPOSURE LIMITS

A. Oral

Humans ingest copper in foods and water. Concentrations in food vary widely from < 10 to > 25,000 µg/100 calories (USEPA 1980). Rich sources are oysters, nuts, liver, kidney, and dried legumes. Earlier estimates of up to 5 mg/day Cu in ordinary diets have been questioned; many diets may contain much less than 1 mg/day (NAS 1980).

A study of copper in the finished water of the 100 largest United States cities found 0.61 to 250 µg Cu/L with a median of 83 µg Cu/L (Durfor and Becker 1964; cited by NAS 1977). There appears to be an increase in copper content of finished water over levels in raw water due to pipe

93

corrosion and chlorination. In a study of the Denver water systems, chlorination resulted in a doubling of the copper content. Finished water from five different source and treatment processes in Denver ranged from 30 to 10.6 μg/L with concentrations at the consumers' taps ranging from 0 to 12 μg/L (Barnett et al. 1969; cited by NAS 1977).

USEPA (1980) gave no criterion for copper in drinking water protective of human health. Undesirable taste and odor may be perceived at > 1 mg Cu/L.

B. Inhalation

Except for a few occupational exposures, daily intakes of airborne copper, even near copper smelters and other major emitters, are neglibible relative to oral ingestion (USEPA 1980).

The threshold limit value (TLV) as a time-weighted average for an 8-h day is 1 mg/m³ for copper dust and mists, with a short-term exposure limit of 2 mg Cu/m³/15 min. The TLV for copper fume is 0.2 mg/m³ (ACGIH 1983). The OSHA permissible exposure limit is 1 mg/m³ for dusts and mists and 0.1 mg/m³ for copper fumes (OSHA 1981).

C. Dermal

Dermal absorption is negligible through intact skin (USEPA 1980).

D. Total Body Burden and Balance Information

According to Snyder et al. (1975), the copper balance for 70-kg reference man in mg/day is: intake 3.5 from food and fluids and 0.02 from air; losses 0.05 from urine, 3.4 from feces, and < 0.5 by other routes. The estimated body burden is 72 mg (65 mg in soft tissue).

III. TOXICOKINETICS

A. Absorption

Copper is actively absorbed in the stomach and duodenum. Typically, about half of a dose will be absorbed, but this can be decreased by competition with zinc and binding by ascorbic acid and other compounds. No data are available on absorption of inhaled copper (USEPA 1980; Stokinger 1981).

B. Distribution

1. General

Absorbed copper is bound by ceruloplasmin and transported in the plasma. The main storage sites are the liver, brain, and muscles. Average blood levels are 1 mg/L. A normal man (70-kg) stores 70 to 120 mg (USEPA 1980; Piscator 1977).

2. Blood

Snyder et al. (1975) estimate that for a 72 mg body burden, 5.6 mg is in whole blood (3.5 mg in plasma and 2.2 mg in red blood cells). Normal whole blood copper concentrations of 243 adults from 19 U.S. cities were 160 to 3,480 µg/L (mean 89 µg/L). Values were abnormally high in two of the cities (Billings, Montana, and El Paso, Texas), which had copper smelters. Workers with metal fume fever working with copper, zinc, or brass had > 1,600 µg Cu/L serum (Stokinger 1981). Henry (1964) states that there is a sex difference, males typically having 700 to 1,400 µg/L serum and females 850 to 1,550 µg/L. The amount in erythrocytes remains constant in any one individual despite variations in serum copper. There is diurnal variation; the day-to-day variation is three to four times higher in females than males. Seven of 20 workers exposed to 0.6 to 1 mg Cu/m^3 (and to Fe, Cd, and Pb) developed a mild, possibly hemolytic, anemia. Plasma copper concentrations in exposed workers and controls were 1,077 ± 40 µg/L and 989 ± 25 µg/L, respectively; corresponding erythrocyte levels were 28.0 ± 2.6 µg/L and 22.5 ± 2.4 µg/L. The difference in erythrocyte copper concentration was significant (Finelli et al. 1981). The analytical methodology of the study was good.

3. Adipose

Snyder et al. (1975) give 3.6 mg (0.24 µg/g) as the copper content in adipose tissue of reference man.

C. Excretion

The major excretion route is the bile. Other minor routes include sweat, urine, and saliva (from which copper may be reabsorbed) (Piscator 1977; EPA 1980).

IV. EFFECTS

A. Acute and Other Short-Term Exposures

Copper sulfate has been used as an emetic (Piscator 1977). Acute ingestion overdoses cause an immediate metallic taste, then epigastric burning, nausea, vomiting, and in more severe cases, diarrhea. There are ulcers and other local damage to the gastrointestinal tract, jaundice (with liver necrosis and biliary stasis), and suppression of urine. Fatal cases often include secondary effects, such as hypertension, shock, and coma. Some cases came from acidic foods (such as fruit juices and carbonated beverages) in copper-lined containers or with copper check valves (e.g., some drink-dispensing machines), which leach the copper from the component (USEPA 1980; Stokinger 1981). Acute copper poisoning also causes hemolytic anemia (Finelli et al. 1981).

Inhaled dusts and fumes cause irritation of the respiratory tract. This may include chronic lung damage resembling silicosis. Allergic contact dermatitis has been reported (Piscator 1977; Stokinger 1981).

B. Chronic Exposure

Chronic exposure may produce "metal fume fever," which is an influenza-like syndrome, with attacks lasting a day or so. Eventually, one may find nasal ulcerations and bleeding.

Except for patients with inborn metabolic disorders, there are few reports of chronic toxicity. Sheep (which appear to be highly susceptible to copper toxicity) receiving excessive copper have a lethal hemolytic anemia, accompanied by severe degeneration in the liver, kidney, and spleen. The latter may be, in part, consequent to the hemolysis (Piscator 1977; Stokinger 1981). Mild, possibly hemolytic, anemia has been observed in workers exposed to copper in the air at levels at or below the TLV (see part III.B.2).

C. Biochemistry

1. Effects on Enzymes

Copper is an essential element. It is part of several of the most vital enzymes, including tyrosinase, cytochrome oxidase, and amine oxidases. It is essential in the incorporation of iron into hemoglobin (Piscator 1977; USEPA 1980).

2. Metabolism

Copper is largely combined with serum albumin and the α-globulin ceruloplasmin, which serve to transport and regulate copper in the body (Stokinger 1981). The metabolism of copper involves the turnover of the copper-containing enzymes. Two inherited diseases reflect disorders of this metabolism. Menkes' disease, apparently an inability to absorb copper, is a copper deficiency. Wilson's disease is the opposite, with excessive accumulation of copper (USEPA 1980).

3. Antagonisms and Synergisms

Copper interacts with a number of other metals. For instance, excess molybdenum causes a copper deficiency (Piscator 1977).

4. Physiological Requirements

Copper is an essential trace element for all vertebrates. The adult Recommended Daily Allowance (RDA) is 2 to 3 mg (NAS 1980).

D. Specific Organs and Systems

The ubiquity of copper-containing enzymes makes organ-specific effects very rare. For instance, Wilson's disease is characterized by hepatic cirrhosis, brain damage and demyelination, kidney defects, and copper deposition in the cornea and other affected organs (Stokinger 1981).

E. Teratogenicity

There is no evidence that excess copper is teratogenic or not (USEPA 1980).

Copper deficiency in the mother animals produces defects in lambs, rats, and guinea pigs (Shepard 1980).

F. Mutagenicity

There is no evidence that copper is mutagenic. It is highly lethal to microorganisms, and is used as such (USEPA 1980).

G. Carcinogenicity

There is no real evidence that copper is carcinogenic (USEPA 1980).

V. REFERENCES

ACGIH. 1983. TLVs® Threshold limit values for chemical substances and physical agents in the work environment with intended changes for 1983-84. Cincinnati, Ohio: American Conference of Governmental Industrial Hygienists.

Finelli VN, Boscolo P, Salimei E, Messineo A, Carelli G. 1981. Anemia in men occupationally exposed to low levels of copper. Heavy Met Environ, Internatl. Conf., 3rd. Edinburgh, UK: CEP Consult. Ltd., pp. 475-478.

Henry. RJ. 1964. Clinical chemistry: principles and technics. New York, NY: Hoeber Medical Division, Harper & Row Publishers.

NAS. 1977. Natl. Academy of Sciences. Drinking water and health. Vol. I. Washington, DC.

NAS. 1980. Natl. Academy of Sciences. Recommended dietary allowances. 9th ed. Washington, DC: Printing and Publishing Office, National Academy of Sciences.

OSHA. 1981. Occupational Safety and Health Admin. Occupational safety and health standards. Subpart 2--Toxic and hazardous substances. Code of federal regulations 29 (Part 1910.1000) pp. 673679.

Piscator M. 1977. Copper. In: Toxicology of metals - Vol. II. Springfield, VA: National Technical Information Service, pp. 206-221, PB 268-324.

Shepard TH. 1980. Catalog of teratogenic agents, 3rd ed. Baltimore: The Johns Hopkins University Press.

Snyder WS, Cook MJ, Nasset ES, Karhausen LR, Howells GP, Tipton IH. 1975. International Commission on Radiological Protection. Report of the task group on reference man. New York. ICRP Publication 23.

Stokinger HE. 1981. Chapter 29. The metals. In: Patty's industrial hygiene and toxicology. 3rd ed. Volume 2A. Clayton GD, Clayton FE, eds. New York: A Wiley-Interscience Publication. John Wiley and Sons, pp. 1493-2060.

USEPA. 1980. U.S. Environmental Protection Agency. Ambient water quality criteria for copper. Springfield, VA: National Technical Information Service, PB 81-117475.

GALLIUM

MAMMALIAN TOXICITY SUMMARY

I. INTRODUCTION

A. Occurrence and Production

Gallium occurs naturally at the rate of 15 g/ton of the earth's crust. It is present in aluminum ores at levels of 0.005 to 0.01%. In small amount, it is found in the minerals sphalerite and germanite. It is also present in tin, iron, manganese, and chromium ores. It is produced as a by-product of zinc and aluminum production (Stokinger 1981).

B. Uses

Gallium arsenide and phosphide are used in light-emitting diodes (LEDs) and the arsenide is used in microwave devices. Fluorescent lamps and superconductors are other uses for gallium compounds and intermetallics.

C. Chemical and Physical Properties

Gallium, atomic no. 31, melts at 29.8°C, and boils at 1700°C. It expands on solidification (Browning 1969; Stokinger 1981). Compounds are normally trivalent; +1 and +2 oxidation states are possible under reducing conditions. Gallium's chemistry is similar to that of aluminum. Ga_2O_3 is slightly soluble in hot acid and alkalies (Stokinger 1981).

II. EXPOSURE AND EXPOSURE LIMITS

A. Oral

No information in secondary sources.

B. Inhalation

No information.

C. Dermal

No information.

D. Total Body Burden and Balance Information

No information.

III. TOXICOKINETICS

A. Absorption

Gallium is not readily absorbed from the gastointestinal tract but is readily taken up by tissue from subcutaneous or intravenous injection (Browning 1969).

B. Distribution

1. General

Gallium is chiefly deposited in bone tissue and is relatively immobile. The gallium content of the bones of rabbits stayed constant for 6 mo. Initially the liver, kidney, and spleen also had high concentrations, but these decreased (Dudley and Marrer 1952; cited by Browning 1969).

2. Blood

Gallium remains in the blood only a short time; it is either excreted or deposited in tissue (Stokinger 1981). Snyder et al. (1975) estimate 16 µg in whole blood of 70-kg reference man and < 1.6 µg in plasma.

C. Excretion

Gallium has been reported to be rapidly excreted especially in the urine. One study found the kidneys able to excrete 50% within 24 h (Browning 1969). Another study reported humans able to excrete 4 to 55% of a total dose within 4 days (Browning 1969).

IV. EFFECTS

A. Acute and Other Short-Term Exposures

Gallium appears to be more toxic to larger than to smaller animals with LD_{50}'s for subcutaneous injection ranging from 600 mg/kg for mice to 10 to 15 mg/kg for dogs and goats and by intravenous injection from > 200 mg/kg for rats to 18 mg/kg for dogs. Signs of toxicity in animals receiving large doses included rapid loss of weight, hyperexcitability, photophobia, blindness, terminal flaccid paralysis, severe gastrointestinal disturbances, anorexia, diarrhea, blood in feces, coma, and death. Tubular damage in the kidneys ranged from cloudy swelling to necrosis (Browning 1969).

Human exposures have involved use of radioactive plus stable gallium in therapeutic doses so toxicity reported may be due to radioactivity. Signs of toxicity included dermatitis, gastrointestinal disturbances, and bone marrow depression (due to the radioactivity) (Browning 1969).

In an industrial case, a woman exposed to GaF_3 fumes reported skin rash and neurological pain and muscle weakness that persisted for 3 mo. The rash cleared in 2 wk (Stokinger 1981).

B. Chronic Exposure

Dogs given 2 to 5 mg/kg intravenously twice a week experienced kidney damage, fatal to some, and mice receiving 40 to 100 mg/kg subcutaneously once a week had degenerative changes in the gastric glands and mucosa (Browning 1969). Exposure duration was not stated.

Rats fed up to 1,000 ppm for 26 wk had no toxic response (Stokinger 1981).

C. Biochemistry

1. Effects on Enzymes

No information.

2. Metabolism

No information.

3. Antagonisms and Synergisms

No information.

4. Physiological Requirements

Gallium is considered necessary for growth of Aspergillus niger but not for green plants or animals (Browning 1969).

D. Specific Organs and Systems

1. Kidneys

Kidneys of rats and mice given gallium showed tubule damage; it was most serious in the rats with uremia, the cause of death at high doses in both species (Browning 1969).

2. Blood

In dogs receiving fatal doses, there were lymph system and blood system changes, lymphadenitis with polymorphonuclear infiltration, aplastic changes in bone marrow, and a 40% fall in hemoglobin levels (Browning 1969).

E. Teratogenicity

Not teratogenic in hamsters when given on the 8th day of pregnancy at 40 mg/kg (Ferm and Carpenter 1970; cited by Shepard 1980).

F. Mutagenicity

No information.

G. Carcinogencity

No information.

V. REFERENCES

Browning E. 1969. Toxicity of industrial metals. 2nd ed. London: Butterworths.

Shepard TH. 1980. Catalog of teratogenic agents, 3rd ed. Baltimore: The Johns Hopkins University Press.

Snyder WS, Cook MJ, Nasset ES, Karhausen LR, Howells GP, Tipton IH. 1975. International Commission on Radiological Protection. Report of the task group on reference man. New York. ICRP Publication 23.

Stokinger HE. 1981. Chapter 29. The metals. In: Patty's industrial hygiene and toxicology. 3rd ed. Volume 2A. Clayton GD, Clayton FE, eds. New York: A Wiley-Interscience Publication. John Wiley and Sons, pp. 1493-2060.

GERMANIUM

MAMMALIAN TOXICITY SUMMARY

I. INTRODUCTION

    A. Occurrence and Production

        Germanium occurs naturally in small concentrations in minerals such
as argyrodite ($4Ag_2S.GeS_2$) and germanite ($7CuS.FeS.GeS_2$), as well as in zinc
ores containing cadmium. It is produced as a by-product of the production of
electrolytic zinc or recovered from flue dusts from the combustion of
producer gas, which may contain up to 3% germanium (Browning 1969).

    B. Uses

        As a semiconductor, germanium is used in the manufacture of
electronic components (transistors, diodes). Is is also used in electroplating metals, in production of metal alloys, in glass lens production, and
as a catalyst in coal hydrogenation (Browning 1969).

    C. Chemical and Physical Properties

        Germanium, atomic no. 32, melts at 937.2°C and boils at 2700°C. It
is insoluble in hydrochloric acid, slowly soluble in sulfuric acid. It forms
salts and hydrides through the action of reducing acids on germanium-containing
ores. Common compounds exhibit +2 and +4 oxidation states resembling stannous
and stannic analogs. GeO_2 is moderately soluble. $GeCl_4$ is hydrolyzed by water.
(Browning 1969; Merck Index 1983).

II. EXPOSURE AND EXPOSURE LIMITS

    A. Oral

        Food contributes about 1.5 mg Ge/day to the daily intake
(Snyder et al. 1975).

    B. Inhalation

        The threshold limit value as a time-weighted average in workroom
air for an 8-h day is 0.6 mg/m³ for germanium tetrahydride; the short-term
exposure limit is 1.8 mg/m³ (ACGIH 1983).

        Because of the presence in coal ash, urban air may contain
germanium (Snyder et al. 1975).

    C. Dermal

        No information.

103

D. Total Body Burden and Balance Information

According to Snyder et al. (1975), the germanium balance for 70-kg reference man is:

Intake, mg/day		Losses, mg/day	
Food and fluids	1.5	Urine	1.4
		Feces	0.10
		Sweat	?

III. TOXICOKINETICS

A. Absorption

Germanium is rapidly absorbed by all routes of administration (Stokinger 1981). Oral absorption is $\geq$ 96% (Synder et al. 1975).

B. Distribution

1. General

Distribution appears to be general, with liver and kidney showing the highest initial concentration (Browning 1969).

2. Blood

Germanium is rapidly moved into the blood and transported to various tissues. Small doses by oral and parenteral routes in animals give blood levels of < 10 µg/g within a few hours. Germanium is cleared from the bloodstream, where it may be transported unbound to protein, within a few hours (Stokinger 1981).

C. Excretion

Germanium is rapidly excreted mostly in the urine. In rabbits and dogs, over 75% was excreted within 72 h (Browning 1969).

IV. EFFECTS

A. Acute and Other Short-Term Exposures

Germanium and GeO_2 are relatively nontoxic by all routes of administration in animal tests. Doses as high as 600 mg/kg given intra-peritoneally have been tolerated without toxic effects in rats, whereas 700 mg/kg has been reported as the minimal lethal dose and 1,200 mg/kg as the absolute lethal dose. In animals, acute toxic effects noted were edematous, hemorrhagic lungs, multiple petechia in the walls of the small intestine, and peritoneal effusion. Death was preceded by hypothermic shock and almost complete respiratory depression (Stokinger 1981). Rats exposed to germanium through inhalation showed no irritation or adverse effects and clearance was rapid (Browning 1969). The inhalation of GeH_4 and $GeCl_4$ did have adverse effects, changes in the respiratory system, nervous system, blood, and kidneys. Alkyl-substituted germanium compounds are not so low in toxicity as inorganic germanium compounds, but they have ∿ 0.1 the toxicity as corresponding organotin and organolead compounds (Stokinger 1981).

B. Chronic Exposure

In a 14-wk study, rats fed 1,000 ppm germanium experienced 50% mortality, rats given 100 ppm in water experienced 50% mortality in 4 wk, but no gross pathological changes were found to account for deaths (Stokinger 1981).

C. Biochemistry

1. Effects on Enzymes

No information.

2. Metabolism

Germanium appears to disturb tissue water balance and metabolism at high exposure levels leading to dehydration and related effects (Stokinger 1981).

3. Antagonisms and Synergisms

No information.

4. Physiological Requirements

Not known to be essential to any living organisms (Stokinger 1981; Browning 1969).

D. Specific Organs and Systems

Systems affected by high doses of germanium compounds included the gastrointestinal system, lungs, liver, and kidneys; but no serious pathological changes have been reported in most cases (Stokinger 1981).

E. Teratogenicity

Not teratogenic in hamsters after i.v. injection of 40 or 100 mg/kg on day 8 of gestation (Ferm and Carpenter 1970; cited by Shepard 1980).

F. Mutagenicity

No information.

G. Carcinogencity

No information.

V. REFERENCES

ACGIH. 1983. TLVs[®] Threshold limit values for chemical substances and
physical agents in the work environment with intended changes for 1983-84.
Cincinnati, Ohio: American Conference of Governmental Industrial Hygienists

Browning E. 1969. Toxicity of Industrial Metals. 2nd ed. London:
Butterworths.

Merck Index, 10th ed. 1983. Rahway, New Jersey: Merck & Co., Inc.

Shepard TH. 1980. Catalog of teratogenic agents, 3rd ed. Baltimore: The
Johns Hopkins University Press.

Snyder WS, Cook MJ, Nasset ES, Karhausen LR, Howells GP, Tipton IH. 1975.
International Commission on Radiological Protection. Report of the task
group on reference man. New York. ICRP Publication 23.

Stokinger HE. 1981. Chapter 29. The metals. In: Patty's industrial
hygiene and toxicology. 3rd ed. Volume 2A. Clayton GD, Clayton FE, eds.
New York: A Wiley-Interscience Publication. John Wiley and Sons,
pp. 1493-2060.

GOLD

MAMMALIAN TOXICITY SUMMARY

I. INTRODUCTION

A. Occurrence and Production

Gold occurs as native gold in veins or in alluvial deposits and in various telluride minerals. Alaska, California, South Dakota, and Utah are principal gold-producing areas in the United States. Gold is recovered from its ores by cyanidation and, less commonly, amalgamation. Gold is also recovered in North America as a by-product of the smelting of ores containing Cu, Pb, Zn, and Ni and purified by electrorefining (Stokinger 1981).

B. Uses

Gold is used in jewelry and arts, dentistry, and electrical and electronic applications. It is used in the construction of printed circuits, microcircuits, integrated circuits, and transistors. In medicine, colloidal gold compounds are used to treat arthritis (Stokinger 1981).

C. Chemical and Physical Properties

Gold, atomic no. 79, melts at 1064°C and boils at 2807°C. It is very unreactive. Gold will dissolve in aqua regia (HNO_3 and HCl) and in cyanide solutions containing oxygen. The latter give the $Au(CN)_2$ complex. $AuCl_3$, $NaAu(CN)_2$, and $NaAuCl_4 \cdot 2H_2O$ are among the water-soluble compounds of gold. All industrially useful gold compounds are trivalent; univalent compounds are also known. In the presence of organic reducing agents, solutions of gold salts give colloidal gold, which may be red, purple, or blue depending on particle size (Stokinger 1981).

II. EXPOSURE AND EXPOSURE LIMITS

A. Oral

Gold alloys are commonly used as dental inlays and crowns. It is doubtful that these materials contribute any gold to the body burden since neither attrition nor leaching is likely.

B. Inhalation

No information.

C. Dermal

Some persons develop eczema from contact with gold jewelry. Photographers may develop dermatitis from contact with $AuCl_3$ (Stokinger 1981)

D.   Total Body Burden and Balance Information

Snyder et al. (1975) estimate a total body burden of < 9.8 mg for 70-kg reference man.

III.   TOXICOKINETICS

A.   Absorption

Stokinger (1981) states that gold is absorbed after ingestion of gold salts, but the degree is not quantitated.

B.   Distribution

1.   General

Practically all the estimates for tissue gold concentrations estimated by Snyder et al. (1975) are given as less than values of 10^{-8} to 10^{-4} g. They estimate that about half of the < 9.8 mg body burden is in skeleton and about half is in total soft tissue. Highest single tissue values given are 0.52 mg in cortical bone and 0.17 mg in adipose tissue. Liver, skin, and muscle may have higher contents than adipose. Stokinger (1981) states that gold is carried in the plasma to practically all parts of the body. More than 75% of an injected dose is ultimately found in kidneys, liver, skin, hair, and other tissues; but quantitation has been difficult even with neutron activation analysis. Concentrations are apparently in the tenths of the parts-per-million range for the tissues mentioned (dry weight). Tissue concentrations are similar in patients with and without toxic symptoms. An autopsy on a patient 8 wk after the end of 4-yr therapy in which the patient had received a total of 2,530 mg Au (as Au thioglucose) revealed highest tissue gold concentrations in organs with abundant reticuloendothelial cells (lymph nodes, liver, sternal bone marrow, and spleen). Much less was in articular and para-articular structures, synovium, bone, cartilage, and muscle.

2.   Blood

Snyder et al. (1975) have estimated the gold content of whole blood as 0.21 µg for 70-kg reference man. Normal blood values reported for adult patients are < 2.5 µg/L. In the patient described in Part III.B.1, who received a total of 2.53 g Au in 4 yr, plasma contained 9.5 mg/L and other biological fluids had similar concentrations. Cellular biologic effects from gold would be expected at such levels. Patients who develop toxicity (nephrotic syndrome or dermatitis) do not show statistically significant differences in their serum gold concentrations (Stokinger 1981).

3.   Adipose

Snyder et al. (1975) estimated 0.17 mg (0.011 µg/g) for the total content of gold in adipose tissue of 70-kg reference man.

C. Excretion

Normal values for gold in urine are $\leq$ 20 µg/L. The patient described in III.A.1 and III.A.2 had 9.9 mg Au/L urine. In 18 patients on gold therapy (Au Na thiomalate), urine contained 23.9 to 30.5% of the dose and feces, 10.3 to 17%. The higher urine, lower fecal values were for the "most improved group." About half of the patients experienced toxic symptoms (mostly dermatitis) (Stokinger 1981).

IV. EFFECTS

A. Acute and Other Short-Term Exposures

Apparently all the lethal animal test results available for gold salts used in arthritis therapy are by injection routes. The results indicate high acute toxicity by these routes for most salts tested; e.g., the i.m. LD_{50} for Na aurothiosulfate is 35 mg/kg in rats (Stokinger 1981). The i.m. injections of Au Na thiomalate and Au thioglucose are painful and produce high initial gold concentrations in the serum. These high levels may damage the kidney. [New oral preparations such as triethylphosphine gold peracetylthioglucose give a lower effective dose without side effects (Science 1981).]

B. Chronic Exposure

Contact with or injections of gold compounds may cause chronic contact sensitivity dermatoses. Skin rashes occur at doses $\geq$ 250 mg Au compound. Tolerance may develop. A delayed hypersensitivity immune reaction may result from skin contact with the crystals of gold compounds. Rarely, a "gold nephrosis" syndrome develops in patients treated with gold compounds. The syndrome characterized by edema of the face and ankles, a red rash around the eyes, and general erythematous plaques (Stokinger 1981).

C. Biochemistry

1. Effects on Enzymes

Au thiomalate inhibits protaglandin synthesis, human epidermal acid phosphatase, and trytophanyl-TRNA synthetase. $AuCl_3$ inhibits the transamidase that synthesizes glucosamine-6-phosphate in connective tissue (Stokinger 1981).

2. Metabolism

No information.

3. Antagonisms and Synergisms

No information.

D. Specific Organs and Systems

1. Kidney

A nephrotic syndrome is rare in patients on gold salt arthritis therapy (Stokinger 1981).

2. Skin

Contact dermatitis is common in patients given gold salts (Stokinger 1981).

3. Blood

In "gold nephrosis," albumin is reduced in the blood; α_2-globulin (marked), β-globulin (slight), and serum cholesterol are increased (Stokinger 1981).

E. Teratogenicity

Gold has been found in a 20-wk fetus of a woman who had received gold therapy. In two reports of 26 and 93 pregnancies among women given gold therapy, two birth defects were noted in the second group. One infant had a dislocated hip and one, a flattened acetabulum. In pregnant rats exposed to Au from aurothiomalate at levels somewhat higher than those of serum during gold therapy, the malformation incidence was 25%. The defects included hydronephrosis, hydrocephalus, and some defects of the eye, heart, and palate. In another study, hydrocephalus, eye defects, and rib fusion occurred in rats and gastroschisis and umbilical hernia occurred in rabbits (Shepard 1980).

F. Mutagenicity

No information.

G. Carcinogenicity

RIII X CBA female mice who had been injected with Au thioglucose showed an increased incidence of mammary cancer (38/38 treated vs. 29/34 untreated) and reduced average time of 50% tumor appearance (240 days vs. 350 days in controls). Four treated vs. zero untreated male mice developed tumors. The treated mice were persistently obese and failed to breed. Since a single injection of 400 mg/kg caused lesions in the brain, especially the hypothalamus, hormonal imbalance was felt to be the mechanism for the mammary cancer increase (Stokinger 1981).

V. REFERENCES

Shepard TH. 1980. Catalog of teratogenic agents, 3rd ed. Baltimore, MD: The Johns Hopkins University Press.

Science.   1981.   New ways to use metals for arthritis.   Science 212:430-431.

Snyder WS, Cook MJ, Nasset ES, Karhausen LR, Howells GP, Tipton IH.   1975. International Commission on Radiological Protection.   Report of the task group on reference man.   New York.   ICRP Publication 23.

Stokinger HE.   1981.   Chapter 29.   The metals.   In: Patty's industrial hygiene and toxicology, 3rd ed., volume IIA, Toxicology.   Clayton GD, Claytor FE, eds.   New York: Wiley-Interscience, John Wiley & Sons, pp. 1493-2060.

HAFNIUM

MAMMALIAN TOXICITY SUMMARY

I. INTRODUCTION

A. Occurrence and Production

The crustal abundance of hafnium is 5 ppm. Hafnium is found in all zirconium minerals. There is usually ∿ 0.5 to 2.0% Hf in zircon and baddeleyite. Hafnium is removed from zirconium only for nuclear applications since Hf has a much greater thermal neutron capture cross section than Zr. Separation methods include solvent extraction, fractional crystallization, and ion exchange (Smith and Carson 1978).

B. Uses

Hafnium is used as a flashbulb filler (Smith and Carson 1978). It is used for control rods in water-cooled reactors, light bulb [flashbulb?] filaments, electrodes, and special glasses and as a getter in vacuum tubes (Hawley 1981).

C. Chemical and Physical Properties

Hafnium, atomic no. 72, melts at 2227°C. The chemical behavior of Hf and Zr is more similar than for any other pair of elements. (Because of the lanthanide contraction, the ionic and atomic radii of Hf and Zr are practically identical [Smith and Carson 1978].) The oxide HfO_2 is water insoluble (Hawley 1981).

II. EXPOSURE AND EXPOSURE LIMITS

See the remarks regarding exposure in the zirconium profile. Hafnium accompanies all exposures to zirconium except any that might be due to the Zr alloys used in the nuclear industry.

A. Oral

No specific information.

B. Inhalation

The threshold limit value as a time-weighted average in workroom air for an 8-h day is 0.5 mg/m³ for hafnium; the short-term exposure limit is 1.5 mg/m³/15 min (ACGIH 1983). The OSHA permissible exposure limit is 0.5 mg/m³ (OSHA 1981).

C. Dermal

No specific information.

D. Total Body Burden and Balance Information

Excreted Hf may range from < 1 to 74 µg Hf/day (Smith and Carson 1978). Snyder et al. (1975) give no body burden or balance information for hafnium.

III. TOXICOKINETICS

A. Absorption

ACGIH (1980) assumes 100% gastrointestinal absorption of $HfOCl_2$ and 10% absorption from the lungs.

B. Distribution

1. General

Four days after an i.v. injection of ^{181}Hf as Na Hf mandelate, the Hf distribution in the rat was 15.4% in bone, 37.8% in liver, 1.5% in kidney, and 5.0% in excreta. Although liver activity was high in the rat after such an injection, little ^{181}Hf was found in the bile duct (Smith and Carson 1978).

2. Blood

No specific information.

C. Excretion

Elimination of an i.v. dose of Hf mandelate in rats was slow (ACGIH 1980). United States astronauts during four missions excreted 0.32 to 4.2 µg Hf/man/day. Four days after an i.v. injection of Na Hf mandelate in rats, 5.0% of the initial Hf dose was in the urine and feces (Smith and Carson 1978)

IV. EFFECTS

A. Acute and Other Short-Term Exposures

The i.p. LD_{50} of $HfOCl_2$ for mice was 76 mg/kg. The i.v. LD_{50} of Hf mandelate for rats was ∼ 100 mg/kg; the LD_{50} for the gluconate was 2,000 to 3,000 mg/kg (ACGIH 1980). Intravenous injections of Hf chloride, nitrate, or acetate were less acutely toxic than injection of the corresponding Zr compounds (Smith and Carson 1978).

B. Chronic Exposure

A borderline toxic response was observed in the livers of most rats fed diets containing 10% Hf as $HfCl_4$ for 90 days. The effect was seen occasionally in rats fed diets with 0.1% Hf (ACGIH 1980).

C. Biochemistry

No information on the biochemistry of hafnium was found in secondary sources.

D. Specific Organs and Systems

See part IV.B.

E. Teratogenicity

No information.

F. Mutagenicity

No information.

G. Carcinogenicity

No information.

V. REFERENCES

ACGIH. 1980. Documentation of the threshold limit values. 4th ed. Cincinnati, Ohio: American Conference of Governmental Industrial Hygienists, Inc.

ACGIH. 1983. TLVs® Threshold limit values for chemical substances and physical agents in the work environment with intended changes for 1983-84. Cincinnati, Ohio: American Conference of Governmental Industrial Hygienists.

Hawley GG. 1981. The Condensed Chemical Dictionary, 10th ed. New York, NY: Van Nostrand Reinhold Co.

OSHA. 1981. Occupational Safety and Health Admin. Occupational safety and health standards. Subpart 2--Toxic and hazardous substances. Code of federal regulations 29 (Part 1910.1000) pp. 673-679.

Smith IC, Carson BL. 1978. Trace Metals in the Environment. Volume 3-Zirconium. Ann Arbor, MI: Ann Arbor Science Publishers, 405 pp.

INDIUM

MAMMALIAN TOXICITY SUMMARY

I. INTRODUCTION

A. Occurrence and Production

The average abundance of indium in the earth's crust is ~ 0.1 ppm. It is present in the major metals in the waste streams produced, or is recovered as a by-product in the smelting and refining of certain sulfide minerals of zinc, iron, and copper and of the tin mineral cassiterite. Indium's ubiquitous presence in urban air and sewage sludges (usually at much lower concentrations than those of other industrial metals) may reflect the fact that indium is a common contaminant of the widely used metal zinc (Smith et al. 1978).

B. Uses

Indium is one of the lowest volume metals consumed in the United States. About half is used in electronics and electrial applications, especially as glass-sealing alloys. Indium antimonide, arsenide, and phosphide are used in infrared detectors and semiconductor applications. Indium-silver alloys are used for brazing; indium-cadmium-silver alloy is used in nuclear reactor control rods. A major use is in low-melting alloys for applications such as meltable safety devices. There are few commercial uses of organoindium compounds; the reaction of $(CH_3)_3In$ with $(C_2H_5)Sb$ produces epitaxial layers of InSb in the electronics industry (Smith et al. 1978).

C. Chemical and Physical Properties

Indium, atomic number 49, is a member of the Group IIIA elements of the periodic table along with B, Al, Ga, and Tl. However, indium's chemistry is not closely analogous to the chemistries of Al and Ga. Compounds corresponding to valences of +1 and +2 have been isolated, but only In(III) compounds are stable in aqueous systems. Indium melts at 156.6°C and boils at 2075°C. Although not attacked by alkalies, indium will dissolve in acetic, oxalic, and cold, concentrated oxidizing mineral acids. Indium(III) salts begin to precipitate as $In(OH)_3$ and/or basic salts at pH 3.4 and above. Neither In or In_2O_3 is appreciably volatile; above ~ 700°C, In_2O predominates.

II. EXPOSURE AND EXPOSURE LIMITS

A. Oral

Drinking water is unlikely to be a major source of human exposure to indium. However, indium might be expected to leach from galvanized iron pipes. No drinking water concentrations have been reported, however. Consumption of fish and shellfish that have bioconcentrated indium from contaminated waters might lead to human oral exposure. The bioconcentration factors, however, are usually low, seldom > 100. Plants and animals usually contain < 0.05 ppm indium, although up to 21 ppm in tissues has been detected in contaminated environments. The daily human intake from food is estimated to be ≦ 8 µg (Smith et al. 1978).

B. Inhalation

Elevated indium levels in ambient air might be expected in lead smelting emissions at El Paso, Texas; Kellogg, Idaho; and East Helena, Montana. Copper smelting emissions in Arizona and at El Paso and Tacoma, Washington and tin smelter emissions in Texas City, Texas are other possible sites of elevated atmospheric indium. Rural and urban sites in the United States and Europe frequently have 0.04 to 0.2 ng In/m^3. Concentrations up to 43 ng/m^3 have been reported near metallurgical operations (Smith et al. 1978).

The threshold limit value as a time-weighted average in workroom air for an 8-h day is 0.1 mg In/m^3 for indium and compounds; the short-term exposure limit is 0.3 mg In/m^3 (ACGIH 1983).

C. Dermal

No information found.

D. Total Body Burden and Balance Information

Indium has been used as an internal standard when multielement analyses of human tissues have been done. No human tissue concentrations were found by Smith et al. (1978) or Fowler (1977) in their reviews.

III. TOXICOKINETICS

A. Absorption

1. Oral

Indium absorption after ingestion is < 0.1 to 2% (Smith et al. 1978).

2. Inhalation

Indium compounds are moderately absorbed after inhalation (Fowler 1977). After one-time inhalation or intratracheal intubation of $In(OH)_3$ or In citrate complex, at least 30% is absorbed within 8 days. The biological half-life for $InCl_3$ in the lungs is < 1 h, but In_2O_3 is absorbed much more slowly. The biological half-life for most indium compounds is about 2 mo (Smith et al. 1978).

3. Dermal

No information.

B. Distribution

1. General

Soluble indium injected at $>$ pH 4 is hydrolyzed to $In(OH)_3$ and behaves like colloidal preparations, which accumulate in the liver, spleen, and other organs of the reticuloendothelial system. Pelt, skeleton, and muscle contain the largest amounts of indium; but kidney, spleen, and liver accumulate the highest concentrations. In baboons and humans, $\leq$ 0.1 to $\leq$ 0.5% of injected ionic indium gets into the fetal circulation (Smith et al. 1978).

2. Blood

At pH's $\leq$ 4.0, intravenously injected ionic indium binds to transferrin of the blood; the complex is absorbed to the surface membrane of both mature and immature erythrocytes.

Gofman et al. (1961; cited by Smith et al.) reported 0.78 µg In/mL of human serum; but the X-ray fluorescence spectrography method apparently had interference from other elements. Wolstenhomme (1964; cited by Smith et al. 1978) did not detect indium (detection limit 0.07 ppm) when analyzing dried blood plasma by spark source mass spectrometry.

C. Excretion

Urinary excretion is the primary route for ionic indium whereas fecal excretion is the major elimination route for colloidal indium. The whole-body biological half-life is about 2 wk for both forms (Fowler 1977). Unless stabilized, water-soluble compounds in the blood are hydrolyzed and the insoluble particles formed are phagocytized (Smith et al. 1978).

IV. EFFECTS

A. Acute and Other Short-Term Exposures

Severe vascular shock lasting from 10 min to 1 h developed in three of 770 patients injected with colloidal ^{113}In for liver scanning (Raiciulescu et al. 1972; cited by Fowler 1977). Generally, soluble and colloidal indium compounds are more toxic than insoluble noncolloidal indium compounds. Laboratory mammal LD_{50}'s of parenterally administered colloidal $In(OH)_3$ or colloidal hydrated indium oxide and water-soluble compounds are usually $\leq$ 13 mg In/kg. LD_{50}'s for water-insoluble compounds such as In_2O_3 and InSb are $\sim$ 400 to 180 In/kg.

Oral and intragastric doses of indium compounds to laboratory mammals exhibit low toxicity; e.g., the LD_{50} for $InCl_3$ is 1,100 mg In/kg. Weight loss; liver, kidney, and lung damage; hind leg paralysis; and convulsions commonly occur with lethal doses.

Rats fed 133 mg In/kg as the sulfate for 72 days showed growth depression, inactivity, ruffled coats, and subacute bronchopneumonia; at 72 mg In/kg, no effect was seen after 21 wk.

Three-week feeding studies of guinea pigs with insoluble indium compounds at 0.1 the LD_{50} caused blood changes and an increase of the amino acid content of the urine (Smith et al. 1978).

B. Chronic Exposures

Rats exposed 4 h daily for 3 mo to 64 mg In_2O_3/m^3 air (53 mg In/m^3) showed marked growth depression; increased lung weight, enlargement of the tracheobronchial lymph nodes, and marked alteration of the alveolar walls. At high enough doses, inactivity, blood changes, and organ damage are sequelae of chronic ingestion of soluble indium compounds.

Rats fed up to 2,000 to 3,000 mg In/kg/day as $InCl_3$ for 3 mo showed anemia and mononuclear cells and interalveolar granular exudate in the lungs. In a similar experiment, In_2O_3 was practically nontoxic (Smith et al. 1978).

Rabbits fed 61 mg In/kg/day as $In(NO_3)_3$ for 4.5 mo showed blood and kidney changes, plethora of internal organs, and sometimes necrotic changes in the liver.

Rats ingesting 0.3 mg In/kg as the citrate or acetate in their drinking water (5 ppm) for life did not exhibit any effect on survival, longevity, or tumor incidence. However, their growth was depressed compared to that of the controls. $InCl_3$ (5 ppm In) in the drinking water of mice had a similar effect (Smith et al. 1978).

There is no evidence that indium has produced ill effects in workers exposed to it or its compounds that could be attributed solely to them since other very toxic elements such as lead and arsenic are present in greater concentrations (Smith et al. 1978).

C. Biochemistry

1. Effects on Enzymes

Indium(III) ions inhibit ferroxidase and isocitric dehydrogenase (Smith et al. 1978).

2. Metabolism

Indium(III) appears to react with the phosphate or carboxyl groups and the protonated amino groups of the phospholipid monolayer of biomembranes. It also reacts with the incompletely esterified phosphate groups of nucleic acids (Smith et al. 1978).

3. Antagonisms and Synergisms

The toxic effects of indium compounds are ameliorated by ferric dextran and Thorotrast (thoria). Pretreating mice with nonlethal doses of indium(III), mercury(II), or manganese(II) salts before giving the mice challenge doses of indium sulfate reduces the toxicity of the indium sulfate. Scandium promotes indium deposition in bone (Smith et al. 1978).

D. Specific Organs and Systems

Most critical organs, including the brain, heart, adrenals, and spleen, are damaged by indium compounds.

1. Kidneys

Intravenous doses of soluble indium compounds damage the kidneys, especially the proximal tubules. Granular dystrophy of the kidney is sometimes noted (Smith et al. 1978).

2. Liver

Dystrophy and/or necrosis in the liver is a common histopathological finding (Smith et al. 1978).

3. Lungs

Lethal parenteral doses of noncolloidal In_2O_3 cause lung edema, necrotizing pneumonia, and respiratory difficulties. An intratracheal dose that killed 36% of the rats within 8 mo caused alveolar membrane proliferation and fibrosis. The lungs were affected even after feeding of soluble indium compounds (Smith et al. 1978).

4. Spleen

Intravenous injections of colloidal hydrated indium oxide produced necrosis in the spleen (Fowler 1977).

5. Skin and Eye

Contact with soluble indium compounds in 5% solution or with metallic indium did not affect intact or abraded skin, but local calcium precipitation occurred after intradermal injections. Soluble salts were strongly corrosive to the conjunctiva of rabbits (Smith et al. 1978).

E. Teratogenicity

Indium nitrate is both teratogenic and embryotoxic. When pregnant golden hamsters were injected intravenously with > 1 mg In/kg as the nitrate on day 8 of gestation and sacrificed on days 12 to 14, the fetuses showed limb malformations, especially fusion or absence of digits. Doses of $\geq$ 2 gm/kg killed all the fetuses; fatal doses for the mother were at least 10 mg In/kg (Smith et al. 1978).

F. Mutagenicity

        Reviews by Smith et al. (1978), Stokinger (1981), and Fowler (1977)
do not mention any mammalian or bacterial evidence for the mutagenicity of
indium.

G. Carcinogenicity

        There have been no adequate studies published on the oncological
action of indium or its salts. No increased incidence of tumors was observed
in ∿ 200 mice fed 5 ppm In^{3+} in their drinking water for life (Smith et al.
1978).

V. REFERENCES

ACGIH. 1983. TLVs[®] threshold limit values for chemical substances and
physical agents in the work environment with intended changes for 1983-84.
Cincinnati, Ohio: American Conference of Governmental Industrial Hygienists.

Fowler BA. 1977. Indium. In: Toxicology of metals. Vol. II. Springfield,
VA: National Technical Information Service, pp. 234-241. PB-268 324.

Smith IC, Carson BL, Hoffmeister F. 1978. Trace metals in the environment.
Vol. V - Indium. Ann Arbor, MI: Ann Arbor Science Publishers, 552 pp.

Stokinger HE. 1981. Chapter 29. The metals. In: Patty's industrial
hygiene and toxicology, 3rd ed., volume IIA, Toxicology. Clayton GD, Clayton
FE, eds. New York: Wiley-Interscience, John Wiley and Sons, pp. 1493-2060.

IRIDIUM

MAMMALIAN TOXICITY SUMMARY

I. INTRODUCTION

A. Occurrence and Production

The platinum-group metals (Pt, Pd, Ru, Rh, Ir, and Os) are recovered from placer deposits of two intergrown alloys of the metals and from sulfide-ore bodies (Ni-Cu, Cu, and Cu-Co sulfides). Principal world sources are the Bushveld Complex of South Africa, the Sudbury District of Canada, and the Norilsk region and Kola peninsula of the USSR. Placer deposits such as those in Alaska are minor sources. Most of the new platinum-group metal recovery in the United States is from copper and gold refining (Smith et al. 1978).

B. Uses

Iridium is used by the petroleum-refining and chemical industries and for electrical applications (Stokinger 1981).

C. Chemical and Physical Properties

The chemistry of iridium is more similar to that of rhodium than of the other platinum-group metals. It is the only ore that can be used unprotected in air at temperatures as high as 2300°C. Even aqua regia does not attack iridium. Iridium compounds exhibit +1, +2, +3, +4, and +6 oxidation states (Stokinger 1981).

II. EXPOSURE AND EXPOSURE LIMITS

A. Oral

No information.

B. Inhalation

No information.

C. Dermal

No information.

D. Total Body Burden and Balance Information

No information.

III. TOXICOKINETICS

A. Absorption

Small amounts of $Na_2{}^{192}IrCl_6$ were absorbed after oral ingestion by laboratory mammals (Stokinger 1981).

B. Distribution

1. General

After injection of $Na_2{}^{192}IrCl_6$, ^{192}Ir levels in spleen, kidney, liver, bone, and testes exceeded those in carcass and muscle (Stokinger 1981).

2. Blood

No information.

3. Adipose

No information.

C. Excretion

No information.

IV. EFFECTS

A. Acute and Other Short-Term Exposures

Stokinger (1981) found acute toxicity information only in an 1826 report. The i.v. LD_{L0} for $IrCl_3$ in dogs was 778 mg/kg. $Ir(NH_3)_3Cl_3$ had an LD_{50} (unspecified route) of 1,500 mg/kg in mice. Only one study on iridium's physiological effects was found in a computerized literature search in October 1984: rabbit aortic strips that had been contracted by adrenergic agonists were relaxed by $K[IrBr_5NO]$ but not by the corresponding chloro derivative (Kruszyna et al. 1980).

B. Chronic Exposure

No information.

C. Biochemistry

1. Effects on Enzymes

No information.

2. Metabolism

No information.

3. Antagonisms and Synergisms

No information.

D. Specific Organs and Systems

No information.

E. Teratogenicity

No information.

F. Mutagenicity

No information.

G. Carcinogencity

No information.

V. REFERENCES

Kruszyna H, Kruszyna R, Hurst J, Smith RP. 1980. Toxicology and pharmacology of some ruthenium compounds: vascular smooth muscle relaxation by
nitrosyl derivatives of ruthenium and iridium. J Toxicol Environ Health
6(4):757-773; Chem Abstr 1980. 93(25):230737u.

NAS. 1977. National Academy of Sciences. Platinum group metals.
Washington, DC: National Academy of Sciences Printing and Publishing Office.

Smith IC, Carson BL, Ferguson TL. 1978. Trace metals in the environment.
Vol. 4 - Palladium and osmium. Ann Arbor, MI: Ann Arbor Science Publishers,
193 pp.

Stokinger HE. 1981. Chapter 29. The metals. In: Patty's industrial
hygiene and toxicology. 3rd ed. Volume 2A. Clayton GD, Clayton FE, eds.
New York: A Wiley-Interscience Publication. John Wiley and Sons,
pp. 1493-2060.

IRON

MAMMALIAN TOXICITY SUMMARY

I. INTRODUCTION

A. Occurrence and Production

Iron is widely distributed in the earth's crust and is the major component of the core. Most ores are complex oxides, which are reduced in a blast furnace (Stokinger 1981).

B. Uses

Iron, or more precisely its carbon alloy, steel, is the backbone of modern civilization, with total production exceeding that of all other metals combined. Besides the structural uses, iron and its compounds are used in pigments, magnetic tapes, catalysts, feeds, disinfectants, tanning, and fuel additives. It has been used therapeutically for 3,500 y, now being limited to therapy of iron-deficiency anemia (Stokinger 1981; Hammond and Beliles 1980).

C. Chemical and Physical Properties

Iron compounds, both ferrous (+2) and ferric (+3) have generally low solubilities in water. The major exceptions are the halides and nitrates. Many coordination complexes exist (Stokinger 1981).

II. EXPOSURE AND EXPOSURE LIMITS

A. Oral

Iron is found in virtually every food with higher concentrations in animal tissues than in plant (Hammond and Beliles 1980). In the United States and Europe, men consume $\sim$ 16 mg/day and women, $\sim$ 12 mg/day. Some drinking waters are high in iron. About 140 µg Fe/day comes from average United States drinking waters (Snyder et al. 1975).

B. Inhalation

The threshold limit value as a time-weighted average in workroom air for an 8-h day is 1 mg Fe/m^3 for soluble iron salts; the short-term exposure limit is 2 mg/m^3 (ACGIH 1983). Iron in urban air would contribute $\sim$ 27 µg/day to the intake (Snyder et al. 1975).

C. Dermal

There is no evidence of specific dermal exposure.

D. Total Body Burden and Balance Information

The total body burden is estimated as 4.2 (Snyder et al. 1975) or 4.5 g (Stokinger 1981). According to Snyder et al. (1975), the iron balance for a 70-kg reference man, in mg/day, is: intake from food and fluids 16 and airborne intake 0.03; losses in urine 0.25, in feces 15, and by other routes 0.51. For the 50-kg reference woman, the values are 12 and 0.03, respectively, for intake and 0.20, 11, and 1.2 for losses (0.6 as menstrual loss).

III. TOXICOKINETICS

A. Absorption

Intestinal absorption of iron, especially ferrous, is a complicated active process. The rate of uptake is inversely related to the state of the body's iron stores (Hammond and Beliles 1980). In adults iron absorption is $\sim$ 30% (Snyder et al. 1975).

B. Distribution

1. General

Estimated normal distribution of the body's iron is: storage (ferritin, hemosiderin, etc., in liver, bone marrow, and spleen), 23.5%; hemoglobin, 72.9%; myoglobin, 3.3%; parenchymal iron (cytochromes, etc.), 0.2%; and transferrin, 0.1% (Stokinger 1981).

2. Blood

Absorbed iron is transported by transferrin, a β-globulin. Normal plasma iron levels are 1,290 µg/L and 1,100 µg/L for men and women, respectively, with the transferrin about one-third saturated. Toxic signs are caused by free iron which appears after the carrier is saturated (Stokinger 1981).

There is as much as fivefold diurnal variation in serum iron, and day-to-day and week-to-week variations can be even larger (Henry (1964).

Snyder et al. (1975) estimate that 2.5 g of the total body burden of 4.2 g is in the blood with 2.4 g in the erythrocytes and 3.6 mg in the plasma.

Workers with iron pneumoconiosis had an average 160 µg Fe/100 g serum compared to 127 µg Fe/100 g in nonexposed workers (Stokinger 1981).

3. Adipose

Adipose tissue of reference man contains 360 mg iron (24 µg/g) (Snyder et al. 1975).

C. Excretion

Usually excess iron appears in the feces because it is not absorbed. Normal losses are through sloughing of intestinal cells and bleeding. Excessive absorbed iron is excreted in the urine (Hammond and Beliles
1980).

IV. EFFECTS

A. Acute and Other Short-Term Exposures

Acute toxicity is all too well-known in children who eat Mommy's
iron tablets. Vomiting is often the first sign; then come gastrointestinal
bleeding, lethargy, restlessness, and perhaps gray cyanosis. After a period
of apparent recovery, signs of pneumonitis and convulsions occur; coma may
occur, as well as jaundice. Gastrointestinal bleeding continues throughout.
If the patient survives 3 or 4 days, complete recovery follows rapidly. The
various effects are mostly due to gastrointestinal irritation and its
sequelae, including hemorrhage and dehydration from failure to drink (Hammond
and Beliles 1980).

B. Chronic Exposure

Chronic excessive iron uptake can lead to hemosiderosis (a generalized increased iron content) or to hemochromatosis (specific histological sites for the hemosiderosis), possibly accompanied by fibrosis. The
condition is relatively benign, but may be accompanied by abnormal glucose
metabolism or increased heart disease. Interpretation of epidemiologic data
has been complicated by the question of whether the hemochromatosis is the
effect of excess iron, per se, or whether it is caused by a derangement of
iron metabolism due to another toxic substance (Hammond and Beliles 1980).

Chronic inhalation of iron fumes leads to mottling of the lungs, a
siderosis that is considered a benign pneumoconiosis, nonfibrotic and not
favorable to tubercle bacilli. However, inhalation of iron oxides with
silica produces the usual silicosis, and hematite inhalation has been reported to produce a progressive, massive fibrosis (Stokinger 1981).

C. Biochemistry

1. Effects on Enzymes

Iron is an essential component of several cofactors, including
hemoglobin and the cytochromes (Stokinger 1981).

2. Metabolism

There is an active, complicated homeostatic mechanism for maintaining proper iron levels in the body (Hammond and Beliles 1980).

3. Antagonisms and Synergisms

There are at least 30 industrial substances capable of producing methemoglobinemia by oxidizing normal Fe^{2+} to Fe^{3+} (Stokinger 1981).

4. Physiological Requirements

Iron is an essential mineral. The recommended daily allowance is 10 mg for men and 18 mg for women (NAS 1980).

D. Specific Organs and Systems

As discussed above, iron is a local irritant to the lung (if inhaled) and, more so, to the gastrointestinal tract.

E. Teratogenicity

No data were found.

F. Mutagenicity

No data were found.

G. Carcinogenicity

Lung tumors in iron ore (hematite) mines may be due to radon inhalation. Massive doses of ferric dextran produced tumors at the injection site in animals. Iron lactate and gluconate gave a small incidence of tumors in animals (Stokinger 1981).

V. REFERENCES

ACGIH. 1983. TLVs® Threshold limit values for chemical substances and physical agents in the work environment with intended changes for 1983-84. Cincinnati, Ohio: American Conference of Governmental Industrial Hygienists.

Hammond PB, Beliles RP. 1980. Metals. In: Casarett and Doull's toxicology, the basic science of poisons, Chapter 17. Doull J, Klaassen CD, Amdur MO, eds. New York: Macmillan, pp. 409-467.

Snyder WS, Cook MJ, Nasset ES, Karhausen LR, Howells GP, Tipton IH. 1975. International Commission on Radiological Protection. Report of the task group on reference man. New York. ICRP Publication 23.

Stokinger HE. 1981. Chapter 29. The metals. In: Patty's industrial hygiene and toxicology. 3rd ed. Volume 2A. Clayton GD, Clayton FE, eds. New York: A Wiley-Interscience Publication. John Wiley and Sons, pp. 1493-2060.

MAMMALIAN TOXICITY SUMMARY

I. INTRODUCTION

A. Occurrence and Production

Lead is widely distributed in a variety of minerals. Those used as ores are galena (PbS) and (much less commonly) anglesite ($PbSO_4$) and cerrusite ($PbCO_3$). Lead is also found in uranium and thorium minerals, as a product of radioactive decay.

The metal is produced by smelting. Secondary lead is recovered from scrap, mostly from automobile storage batteries (USEPA 1980; Stokinger 1981).

B. Uses

Most lead is used as the metal and the sulfate in storage batteries. Alloys are used as cable sheaths, as solder, as type metal, as bearings (babbitt metal, etc.), as bullets, as radiation shielding, and in other products. The use of many compounds, such as inorganic pigments and tetra-ethyllead for gasoline, is decreasing. Some new, increasing uses include litharge (PbO) in ceramic magnets (Stokinger 1981).

C. Chemistry

Most inorganic salts are in the +2 oxidation state; some are +4. Water solubility is poor to negligible, except for the acetate, nitrate, and chloride (Stokinger 1981).

II. EXPOSURE AND EXPOSURE LIMITS

The literature on exposure to lead is vast. Two recent reviews, Barltrop (1979) and Biddle (1982), give general overviews of the manifold sources and controversies.

A. Oral

Water-borne lead is minimal, because it forms essentially insoluble sulfates and carbonates. An exception is lead-lined tanks and lead pipes in water distribution systems, which produce toxic concentrations, especially in acid, soft water (Tsuchiya 1977; USEPA 1980).

In a study of finished-water in the 100 largest U.S. cities, the median lead level was 3.7 µg/L with maximum 62 µg/L (Shapiro 1962; cited by NAS 1977). In another similar study, lead was found in 18.1% of finished waters with a mean level of 33.9 µg/L, maximum 139 µg/L (Kopp and Kroner 1967; cited by NAS 1977). McCabe (1970; cited by NAS 1977) in a study of

969 U.S. water systems found an average lead concentration at the tap of 13.1 µg/L; 1.4% of the samples exceeded the drinking water standard of 50 µg/L. Craun and McCabe (1975; cited by NAS 1977) found 95% of the tap-water samples in Seattle exceeded 50 µg/L; in Boston, 65% exceeded the limit. Lead levels at the tap may be higher than in the finished water at the treatment plant; a study of cities with wide usage of lead pipes found mean tap concentrations of lead to be 30 µg/L with 26.7% exceeding 50 µg/L (Karalekas et al. 1975; cited by NAS 1977). The national interim primary drinking water standard for lead is 50 µg/L (USEPA 1975), the same as the 1980 criterion (USEPA 1980).

Therefore, major oral intake comes from food and (in certain cases) ingested dust. Significant amounts may come from lead-based solder in cans. Total ingestion has been estimated at 200 to 300 µg/day. Most toxicity comes from grossly contaminated sources, such as moonshine condensed in multiply-soldered automobile radiators and children with pica (mouthing or eating nonedible objects) in areas with much lead-based paint -- particularly old, peeling paint (USEPA 1980).

B. Inhalation

Ambient air in cities ranges from 0.02 to 10 µg Pb/m^3, coming from automobile exhausts and various industrial sources. Baseline (mid-ocean values) are about 0.001 µg/m^3 (Tsuchiya 1977; USEPA 1980).

The threshold limit value as a time-weighted average in workroom air for an 8-h day is 0.15 mg/m^3 for lead, inorganic dusts and fumes; the short-term exposure limit is 0.45 mg/m^3 for 15 min (ACGIH 1983). The OSHA permissible exposure limit is 0.2 mg/m^2 (OSHA 1981), but NIOSH (1978) has recommended < 0.1 mg/m^3 as the limit.

C. Dermal

Lead naphthenate and other salts of organic acids are slightly absorbed through the skin (Lauwerys 1983).

D. Total Body Burden and Balance Information

According to Snyder et al. (1975), the lead balance for a 70-kg reference man in mg/day is: intake 0.44 from food and fluids and 0.01 from air; losses 0.045 from urine, 0.3 from feces, and 0.1 by other routes. The total burden has been estimated as 100 to 400 mg. It increases with age (Tsuchiya 1977). Snyder et al. (1975) used 120 mg as the total body burden with 11 mg in soft tissues.

III. TOXICOKINETICS

A. Absorption

Normal adults will absorb about 10% of an oral dose of lead compounds. This increases in children (up to 50%), if taken on an empty stomach.

and with some unusual diets. It is reduced for lead organic salts incorporated in dried paint films. Absorption of inhaled lead is more difficult
to generalize about; particles may be deposited in the alveoli and slowly
leach away (Tsuchiya 1977; USEPA 1980). Lauwerys (1983) states that ~ 50%
of lead deposited in the lung is absorbed.

B. Distribution

1. General

Some 90 to 95% of the body lead is deposited in the mineral matrix
of the skeleton. The rest is generally distributed.

2. Blood

The lead concentration in erythrocytes is about 16 times that in
plasma (Tsuchiya 1977; USEPA 1980). Snyder et al. (1975) estimated 1.4 mg Pb
in whole blood, 1.2 mg in erythrocytes, and 0.14 mg in plasma in 70-kg reference man.

Lauwerys (1983) summarized various biological indexes of lead
exposure. His discussion of lead in blood but not the other indexes will be
considered here.

Lead in blood is the best indicator of recent lead exposure in a
steady-state situation, but it does not correlate well with total body burden.
At levels < 10 to 20 µg Pb/m^3 air (nonoccupational exposure), every increase
of 1 µg Pb/m^3 contributes 10 to 20 µg Pb/L to the whole blood. Half of nonoccupationally exposed persons have blood lead concentrations < 200 µg/L; the
95 percentile value is < 350 µg/L. Levels of 600 to 700 µg/L in male lead
workers are usually considered acceptable, but subclinical effects (reduction
in nerve conduction velocity, psychomotor disturbances, and an increased
prevalence of subjective complaints) are detected at blood levels > 500 µg/L.
The World Health Organization in 1980 proposed 400 µg/L for male workers and
300 µg/L for women of child-bearing age as maximal tolerable lead in blood
concentrations. Early effects on heme synthesis as reflected by higher FEP
(free erythrocyte protoporphyrin) and lower ALA-d (δ-aminolevulinic acid
dehydratase) levels than these levels in controls despite normal hemoglobin
and hematocrit values were seen in a group of workers whose blood lead concentrations never exceeded 800 µg/L (mean 480 µg/L; range 330 to 710 µg/L)
(Valentine et al. 1982). (Atomic absorption spectrophotometry analytical
methodology was excellent.)

The cube root of the lead intake from food and water was related to
blood levels in a population exposed to lead in their drinking water. At
100 µg/L, diet and drinking water contributed about the same to the total
week's intake. FAO/WHO has recommended a provisional tolerable intake for
lead by adults as 3 mg/wk. No more than 50% of a population's blood lead
values should exceed 200 µg/L. Using the cube root relationship they developed, Sherlock et al. (1982) calculated that 1.45 mg Pb/wk would give a mean
of 200 µg Pb/L (values determined by atomic absorption spectrophotometry,
excellent quality control). More recently Delves et al. (1984) concluded

that a single blood Pb concentration from an adult is an excellent biological indicator for exposure provided that the degree of exposure has not changed.

3. Adipose

Snyder et al. (1975) estimated 0.6 mg Pb in adipose tissue of reference man (0.04 µg/g).

C. Excretion

Most excretion occurs in the urine. Some occurs in gastrointestinal secretion, hair, sweat, etc. (Tsuchiya 1977).

The half-life of lead in blood and some rapidly exchanging tissues (∿ 2 mg in nonoccupationally exposed subjects) is ∿ 35 days; in soft tissues (0.3 to 0.9 mg in controls), the half-life is ∿ 40 days; and in bone, the half-life is ∿ 20 yr (Lauwerys 1983).

IV. EFFECTS

A. Acute and Other Short-Term Exposures

Acute toxicity is quite unusual, since lead is a relatively insoluble, cumulative poison. Reported signs are fatigue, disturbance of sleep, and constipation, followed by colic, anemia, and neuritis. Occasionally, acute lead encephalopathy, characterized by vomiting, apathy, drowsiness, stupor, ataxia, hyperactivity, seizures, and other neurological signs and symptoms is seen (Stokinger 1981; Tsuchiya 1977).

B. Chronic Exposure

The classic effects of chronic lead poisoning are loss of appetite, metallic taste, constipation and obstipation, anemia, pallor, malaise, weakness, insomnia, headache, nervous irritability, muscle and joint pains, fine tremors, encephalopathy and colic. Some cases develop weakness in the extensor muscles, which is seen as "wrist drop" or "foot drop" (NIOSH 1978). A recent review, with presentation of some new cases is Cullen et al. (1983).

A number of workers have tried to correlate body lead levels with a toxic threshold. The current "maximum acceptable blood lead level" is 30 µg/dL (Biddle 1982). An alternative clinical test is the free erythrocyte protoporphyrin (FEP) level, which is elevated (above 60 µg/dL packed erythrocytes) in lead intoxication (Cullen et al. 1983; Day and Tenant 1982). See also discussion in Part II.B.2.

C. Biochemistry

1. Effects on Enzymes

Lead is an inhibitor of many enzymes. This is the basis of most of the specific toxic effects discussed below (Tsuchiya 1977; Cullen et al. 1983)

2. Metabolism

Other than its deposition in bone and toxic reactions, lead has minimal involvement with metabolism (Tsuchiya 1977; Stokinger 1981).

3. Antagonisms and Synergisms

Lead interacts with a number of other metals. Detoxification is usually by chelation of the lead with calcium ethylenediaminetetraacetate given parenterally. Repeated treatments leach out the bone depots (Tsuchiya 1977).

D. Specific Organs and Systems

1. Erythrocytes

The erythrocytes are the critical organ for toxicity--the one affected at the lowest doses. This was confirmed recently in a study of asymptomatic persons exposed to automobile emissions (Day and Tennant 1982). The toxic mechanism involves inhibition of ferrochelatase (heme synthetase), erythrocyte 5'-nucleotidase, δ-aminolevulinic acid dehydratase, and other enzymes, including some involved in iron transport. These result in decreased erythropoiesis, as well as decreased erythrocyte survival, causing anemia and pallor. Other effects, useful for diagnosis, include reticulocytosis, basophilic stippling of the erythrocytes, increased aminolevulinic acid, coproporphyrin in urine, and increased free erythrocyte porphyrin (protoporphyrin IX) (Tsuchiya 1977; Stokinger 1981; Cullen et al. 1983).

2. Gastrointestinal System

Lead colic was well described by Hippocrates; we have added little to what he knew, except that it normally occurs at lead levels greater than those which cause erythrocyte effects. In addition, some studies have found hepatotoxicity (Tsuchiya 1977; Cullen et al. 1983).

3. Rheumatological Effects

"Lead arthralgia" (joint pains) has long been described. Recent evidence implies that this is actually lead-induced gout caused by lead's interference with uric acid excretion by the kidney. Lead inhibition of guanine aminohydrolase may also be involved (Cullen et al. 1983).

4. Kidney

Nephrotoxicity is most commonly seen in childhood lead intoxication and in moonshine plumbism, which may be more acute than chronic. Lead causes tubular dysfunction, with ultrastructural changes in the mitochondria, which result in amino aciduria, glycosuria, and phosphaturia, as well as gout. Despite earlier reports, there is no apparent increase in hypertension (Tsuchiya 1977; Stokinger 1981; Cullen et al. 1983).

5. Endocrine and Reproductive Effects

Lead is associated with depression of many endocrine functions, particularly the thyroid and adrenal, but details are not clear (Tsuchiya 1977; Cullen et al. 1983).

There is limited, but graphic, evidence that lead is associated with a high incidence of premature deliveries and spontaneous abortions in humans. Lead also suppresses testicular function, even to sterility, in men (Cullen et al. 1983).

6. Nervous System

Lead encephalopathy is well-known in acute lead intoxication. A variety of milder, nonspecific, often "subclinical" effects have been reported in chronic lead intoxication. These include fatigue, irritability, insomnia, nervousness, headache, weakness, loss of appetite, difficulty in sleeping, and abnormal responses in various neuropsychological tests. Particularly controversial are the claims of some that hyperkinetic or aggressive behavior and mental retardation are associated with chronic, low-level lead intoxication (Tsuchiya 1977; Cullen et al. 1983; USEPA 1980).

Peripheral neuropathy involves the motor nerves, but not the sensory. This is seen as decreased nerve-conduction velocity and as motor weakness in the most used muscle groups. There are occasional reports (possibly coincidental) of a syndrome resembling amyotrophic lateral sclerosis. Animal studies have found axonal degeneration and demyelination (Tsuchiya 1977; Stokinger 1981; Cullen et al. 1983).

E. Teratogenicity

Despite fetotoxicity, there is no evidence that lead is teratogenic (Tsuchiya 1977; USEPA 1980). Some reports of teratogenicity, however, were summarized by Shepard (1980).

F. Mutagenicity

Lead intoxication effects include chromosome aberrations (Tsuchiya 1977; Forni et al. 1980).

G. Carcinogencity

There is no evidence that lead is carcinogenic to man. However, lead is carcinogenic, in high doses, in animals; the kidney is most affected (Tsuchiya 1977; USEPA 1980; Stokinger 1981).

V. REFERENCES

ACGIH. 1983. TLVs[®] threshold limit values for chemical substances and physical agents in the work environment with intended changes for 1983-84. Cincinnati, Ohio: American Conference of Governmental Industrial Hygienists.

Barltrop D. 1979. Geochemical and man-made sources of lead and human health. Philos Trans B: Biol Sci R Soc Lond 288:205-211.

Biddle GN. 1982. Toxicology of lead: primer for analytical chemists. J Assoc Off Anal Chem 65(4):947-952.

Cullen MR, Robins JM, Eskenazi B. 1983. Adult inorganic lead intoxication: presentation of 31 new cases and a review of recent advances in the literature. Medicine 62(4):221-247.

Day CM, Tennant FS Jr. 1982. Peripheral blood findings associated with asymptomatic lead exposure. Am J Med Technol 48(2):139-140.

Delves HT, Sherlock JC, Quinn MJ. 1984. Temporal stability of blood lead concentrations in adults exposed only to environmental lead. Hum Toxicol 3(4):279-288; Chem Abstr 1984. 101:145530m.

Forni A, Sciame A, Pertazzi PA, Alessio L. 1980. Chromosome and biochemical studies in women occupationally exposed to lead. Arch Environ Health 35(3):139-146.

Lauwerys RR. 1983. Chapter II. Biological monitoring of exposure to inorganic and organometallic substances. In: Industrial chemical exposure: guidelines for biological monitoring. Davis, CA: Biomedical Publications, pp. 9-50.

NAS. 1977. National Academy of Sciences. Drinking water and health. Vol. I. Washington, DC.

NIOSH. 1978. Natl. Inst. Occupational Safety and Health. Criteria for a recommended standard: occupational exposure to inorganic lead. Washington, DC: U.S. Government Printing Office, DHEW (NIOSH) Publication No. 78-158.

OSHA. 1981. Occupational Safety and Health Admin. Occupational safety and health standards. Subpart 2--Toxic and hazardous substances. Code of federal regulations 29 (Part 1910.1000) pp. 673-679.

Shepard TH. 1980. Catalog of teratogenic agents, 3rd ed. Baltimore: The Johns Hopkins University Press.

Sherlock J, Smart G, Forbes GI et al. 1982. Assessment of lead intakes and dose-response for a population in Ayr exposed to a plumbosolvent water supply. Hum Toxicol 1(2):115-122.

Snyder WS, Cook MJ, Nasset ES, Karhausen LR, Howells GP, Tipton IH. 1975. International Commission on Radiological Protection. Report of the task group on reference man. New York. ICRP Publication 23.

Stokinger HE. 1981. Chapter 29. The metals. In: Patty's industrial hygiene and toxicology. 3rd ed. Volume 2A. Clayton GD, Clayton FE, eds. New York: A Wiley-Interscience Publication. John Wiley and Sons, pp. 1493-2060.

Tsuchiya K. 1977. Lead. In: Toxicology of metals - Vol. II. Springfield, VA: National Technical Information Service, pp. 242-300, PB 268-324.

USEPA. 1975. U.S. Environmental Protection Agency. National interim primary drinking water regulations. (40 FR 59566).

USEPA. 1980. U.S. Environmental Protection Agency. Ambient water quality criteria for lead. Springfield, VA: National Technical Information Service. PB 81-117681.

Valentine JL, Baloh RW, Browdy BL et al. 1982. Subclinical effects of chronic increased lead absorption--a prospective study. Part IV. Evaluation of heme synthesis effects. J Occup Med 24(2):120-125.

LITHIUM

MAMMALIAN TOXICITY SUMMARY

I. INTRODUCTION

    A. Occurrence and Production

        Lithium is widely distributed, principally in aluminum silicates.
It is found in sea water at 11 ppm and at higher levels in some springs.
Commerical production is from spodumene ($LiAlSi_2O_6$) and from lithium carbonate in a Nevada brine well. The ore (or spring salt) is converted to pure
lithium chloride, which is electrolyzed to the metal (Stokinger 1981).

    B. Uses

        Lithium is the lightest metal. Lithium and compounds are used in
the grease and ceramics industries, in metallurgy, air conditioning, and the
chemical industry. Lithium carbonate is used therapeutically as an anti-
depressant (Stokinger 1981; Hammond and Beliles 1980).

    C. Chemical and Physical Properties

        Lithium forms a wide variety of univalent compounds. Lithium
hydride and $LiAlH_4$ decompose in water; most other common compounds are highly
water soluble (Stokinger 1981).

II. EXPOSURE AND EXPOSURE LIMITS

    A. Oral

        Lithium is widely found in plant and animal tissues. Daily intake
has been estimated at 2 mg (Hammond and Beliles 1980); therapeutic doses are
90 to 1,800 mg/day (Shepard 1980).

    B. Inhalation

        No data were found.

    C. Dermal

        No data were found.

    D. Total Body Burden and Balance Information

        Snyder et al. (1975) estimate a burden of 670 µg Li in soft tissues
of reference man. They also give the following estimate for the lithium
balance for the 70-kg reference man (in mg/day): intake from food and fluids
2.0; losses in urine 0.8, in feces 1.2, and traces by other routes (hair,
nails, milk, sweat).

III. TOXICOKINETICS

A. Absorption

Lithium is readily absorbed from the gastroinestinal tract (Hammond and Beliles 1980).

B. Distribution

1. General

Distribution among organs is almost uniform. The greater part is intracellular. Lithium tends to compete with sodium (Hammond and Beliles 1980).

2. Blood

Normal plasma concentration is $\sim$ 17 µg Li/L (Hammond and Beliles 1980). Plasma concentrations of subjects living in Chile increased with increased exposure to lithium in the drinking water, but the increase in the plasma did not parallel the increase in the water (Zaldivar 1980).

C. Excretion

Excretion is principally in the urine, with some in the feces (Hammond and Beliles 1980).

IV. EFFECTS

A. Acute and Other Short-Term Exposures

When a patient is first dosed with lithium carbonate, he often has nausea, vomiting, and abdominal pain about an hour after each dose (at the time of peak plasma level); these symptoms soon disappear (Stokinger 1981). Lithium hydride is highly corrosive to all tissues it contacts, but this is due to the hydride moiety's reacting with water.

B. Chronic Exposure

Chronic toxicity presents a vague, ill-defined picture. The main effects are usually on the gastrointestinal tract, the central nervous system (CNS), and the kidney. Besides the acute signs, thirst increases, and salivation and diarrhea may occur. CNS effects include tremors (especially of the hands), muscular weakness, ataxia, giddiness, tinnitis, drowsiness, muscular hyperirritability and fasciculations, lethargy, stupor, and in extreme cases, coma and seizures. Renal signs include polyuria, elevation of nonprotein nitrogen, and, in the terminal stages, oliguria. Dermatologic reactions (folliculitis) have also been reported (Hammond and Beliles 1980; Stokinger 1981).

Lithium in Houston, Texas, water supplies was positively correlated (statistically significant at the 0.05 probability level) with age-adjusted death rates from arteriosclerotic heart disease (Valentine and Chambers 1976; cited by Zaldivar 1980).

C. Biochemistry

1. Effects on Enzymes

No specific reports.

2. Metabolism

Lithium is not metabolized.

3. Antagonisms and Synergisms

Most of the above mentioned toxic effects are due to competition of lithium ions with those of sodium and/or potassium (Hammond and Beliles 1980; Stokinger 1981). Lithium competes with sodium for renal tubular reabsorption; the toxicity of lithium salts is related to sodium depletion (Snyder et al. 1975).

D. Specific Organs and Systems

As stated above (Section IV.C.3), most effects are due to competition at the molecular level.

E. Teratogenicity

Lithium has been found teratogenic in the rat and mouse. Negative results were reported in the rat (another strain), monkey, and rabbit (Stokinger 1981). An increase in a rare cardiac defect, Ebstein's anomaly, has been noted in the children of women dosed therapeutically with lithium (Shepard 1980).

F. Mutagenicity

No information.

G. Carcinogenicity

No evidence of carcinogenicity.

V. REFERENCES

Hammond PB, Beliles RP. 1980. Chapter 17. Metals. In: Casarett and Doull's toxiciology, the basic science of poisons. Doull J, Klaassen CD, Amdur MO, eds., New York, NY: Macmillan, pp. 409-467.

Shepard TH.   1980.   Catalog of teratogenic agents, 3rd ed.   Baltimore: The Johns Hopkins University Press.

Snyder WS, Cook MJ, Nasset ES, Karhausen LR, Howells GP, Tipton IH.   1975. International Commission on Radiological Protection.   Report of the task group on reference man.   New York.   ICRP Publication 23.

Stokinger HE.   1981.   Chapter 29.   The metals.   In: Patty's industrial hygiene and toxicology.   3rd ed.   Volume 2A, Toxicology.   Clayton GD, Clayton FE, eds.   New York: A Wiley-Interscience Publication.   John Wiley and Sons, pp. 1493-2060.

Zaldivar R.   1980.   High lithium concentrations in drinking water and plasma of exposed subjects.   Arch Toxicol 46:319-320.

MAGNESIUM

MAMMALIAN TOXICITY SUMMARY

I. INTRODUCTION

A. Occurrence and Production

Magnesium is a widely distributed component of the earth's crust. The major commerical sources are magnesite ($MgCo_3$), brucite ($MgO \cdot H_2O$), dolomite ($MgCO_3 \cdot CaCO_3$), and seawater (0.13% Mg). The primary commerical products are the metal, produced by electrolysis or by heating with ferrosilicon, and magnesia (MgO), produced by heating the carbonate (Stokinger 1981).

B. Uses

The metal is used structurally and in alloys, as well as for cathodic protection of iron and steel. Compounds are used as refractories, in cement, as insulation, in chemical processing, as a pharmaceutical, and for other purposes (Stokinger 1981). Magnesium oxide and the trisilicate are used in humans as antacids; the sulfate is used as a cathartic and anticonvulsant (Merck Index 1983).

C. Chemical and Physical Properties

Magnesium is the lightest structural metal, with use limited by its cost and flammability. Its divalent compounds are typical of the alkaline earths (Stokinger 1981).

II. EXPOSURE AND EXPOSURE LIMITS

A. Oral

Most foods contain significant amounts of magnesium. Nuts, cereals, seafoods, and meats are particularly high (Hammond and Beliles 1980). Magnesium was found in the finished water of the 100 largest U.S. cities at a median concentration of 6.25 mg/L, with a range from zero to 120 mg/L (Durfor and Becker 1964; cited by NAS 1977). Up to 20% of the daily intake can be contributed by drinking very hard waters (Crounse et al. 1983). The World Health Organization recommends a limit of 150 mg Mg/L (NAS 1980). Certain medicines can also contribute to oral intake of magnesium.

B. Inhalation

No data are available in the secondary references consulted.

C. Dermal

The general population comes in dermal contact with magnesium in tap water, especially the very hard waters.

D. Total Body Burden and Balance Information

According to Snyder et al. (1975), the magnesium balance for the 70-kg reference man in g/day is: intake from food and fluids 0.34; losses in urine 0.13, in feces 0.21, and by other routes < 0.002. In the 50-kg reference woman, the figures are 0.27, 0.11, 0.16, and < 0.002, respectively. Body burden is estimated at 10 (Snyder et al. 1975) or 20 g (Hammond and Beliles 1980).

III. TOXICOKINETICS

A. Absorption

Absorption is homeostatically regulated (Crounse et al. 1983). Humans absorb ~30% of the dietary intake. High protein in the diet will increase absorption to 40%. Vitamin D, calcium, alcohol, antibiotics, and growth hormone also affect absorption (Snyder et al. 1975).

B. Distribution

1. General

Most of the body's magnesium is in the bones (where concentration is inversely related to that of calcium) and muscle. Magnesium is the second (to potassium) most abundant intracellular cation, at 6 to 20 m equiv/L (Stokinger 1981).

2. Blood

Snyder et al. (1975) estimated 210 mg in whole blood with 68 mg in plasma and 130 mg in erythrocytes. Homeostatic regulation of gastrointestinal absorption and renal excretion maintain constant serum levels of 1.809 ± 0.132 m equiv/L (ultrafilterable content 1.175 ± 0.096 m equiv/L) (Crounse et al. 1983). Thus, blood levels are not a good indicator of body stores or long-term exposure.

Healthy persons ingesting an antacid containing 3.5 g Mg/day showed only a slight serum Mg^{2+} elevation from the control value of 1.6 to 1.7 m equiv/L. Apparently, little of the antacid was being absorbed since urinary Mg^{2+} excretion increased little (from 7.82 to 8.43 m equiv/day). [Serum Mg^{2+} of patients with impaired renal function, however, needs to be monitored closely since Mg^{2+} in serum increased more and rose more rapidly (Lembcke and Fuchs 1984).]

3. Adipose

Reference man is estimated to contain 300 mg Mg (20 µg/g) in adipose tissue (Snyder et al. 1975).

C. Excretion

Magnesium is excreted in the feces (via bile) and urine, with traces in sweat (Hammond and Beliles 1980). Most fecal excretion represents unabsorbed magnesium (Synder et al. 1975).

IV. EFFECTS

A. Acute and Other Short-Term Exposures

Inhalation of magnesium may cause "metal fume fever" and an associated leukocytosis. The syndrome is associated with fresh MgO fumes and consists of fever and leukocytosis, generally brief (a day or less). Subcutaneous magnesium metal or alloy to animals produces a gas gangrene-like response, with massive, localized gaseous tumors with extensive necrosis. This is apparently due ot an inflammatory reaction. In workers, irritation and slowly healing wounds are reported, but nothing like gas gangrene (Hammond and Beliles 1980; Stokinger 1981).

Oral doses of magnesium sulfate are poorly absorbed; osmotic withdrawal of water from the gut wall leads to their purgative effect. Very high Mg dietary levels (1.5 to 2.5%) or increased Mg absorption can cause fatal poisoning (Stokinger 1981).

Intravenous doses of magnesium salt produce local and general anesthesia, narcosis (counteracted by Ca^{2+} i.v.), and muscular paralysis. Cardiac conduction changes are seen in dogs with serum levels of 5 to 10 m equiv/L (Stokinger 1981).

In pregnant women given i.v. $MgSO_4$ to suppress premature labor, serum levels rose to 61 mg/L within 30 min and remained elevated. The hypermagnesemia apparently rapidly decreased the secretion of parathyroid hormone, which remained depressed despite an accompanying reduction in calcium blood levels. No other toxic signs or symptoms were mentioned in the abstract.

B. Chronic Exposure

No serious chronic effects have been reported, except the deficiency syndrome. New users of drinking waters high in magnesium (500 to 1,000 mg/L) may experience a cathartic effect, but they usually become tolerant (NAS 1980). A usual therapeutic dose of $Mg(OH)_2$ is $\geq$ 250 mg as MgO.

C. Biochemistry

1. Effects on Enzymes

Magnesium is essential for neuromuscular conduction. It is involved in many vital enzymes, including phosphatases, ATPases, carbohydrate metabolism, oxidative phosphorylation, etc. (Stokinger 1981).

2. Metabolism

Magnesium is not metabolized.

3. Antagonisms and Synergisms

Magnesium is often competitive with calcium (Hammond and Beliles 1980).

4. Physiological Requirements

Magnesium is an essential mineral. The recommended daily allowance is 350 mg for men and 300 mg for women (NAS 1980).

D. Specific Organs and Systems

Due to its ubiquity in enzymatic reactions, magnesium effects are not localized.

E. Teratogenicity

No teratogenicity studies have been found for simple magnesium salts. Magnesium deficiency has produced defects in rat fetuses (Hurley et al. 1971; cited by Shepard 1980).

F. Mutagenicity

No studies have been found.

G. Carcinogencity

No studies have been found.

V. REFERENCES

Cholst IN, Steinberg SF, Trapper PJ, Fox HE, Segre GV, Bilezikian JP. 1984. The influence of hypermagnesemia on serum calcium and parathyroid hormone levels in human subjects. N Engl J Med 310(19):1221-1225; Chem Abstr 1984. 101:17305e.

Crounse RG, Pories WJ, Bray JT, Mauger RL. 1983. Geochemistry and man: Health and disease 1. Essential elements. In: Applied environmental geochemistry. Thorntain, I., ed., London, UK: Academic Press, pp. 309-333.

Hammond PB, Beliles RP. 1980. Metals. Chapter 17. In: Casarett and Doull's Toxicology, the basic science of poisons. Doull J, Klaassen CD, Amdur MO, eds. New York: Macmillan, pp. 409-467.

Lembcke B, Fuchs C. 1984. Magnesium load induced by ingestion of magnesium-containing antacids. Contrib Nephrol 38 (Trace Elem. Renal Insufficiency): 185-194, Chem Abstr 1984. 101:48474p.

Merck Index, 10th ed. 1983. Rahway, New Jersey: Merck & Co., Inc.

NAS. 1977. National Academy of Sciences. Drinking water and health. Vol. I. Washington, DC.

NAS. 1980. National Academy of Sciences. Recommended dietary allowances. 9th ed. Washington, DC: Printing and Publishing Office, National Academy of Sciences.

Shepard TH. 1980. Catalog of teratogenic agents, 3rd ed. Baltimore: The Johns Hopkins University Press.

Snyder WS, Cook MJ, Nasset ES, Karhausen LR, Howells GP, Tipton IH. 1975. International Commission on Radiological Protection. Report of the task group on reference man. New York. ICRP Publication 23.

Stokinger HE. 1981. Chapter 29. The metals. In: Patty's industrial hygiene and toxicology. 3rd ed. Volume 2A. Clayton GD, Clayton FE, eds. New York: A Wiley-Interscience Publication. John Wiley and Sons, pp. 1493-2060.

MANGANESE

MAMMALIAN TOXICITY SUMMARY

I. INTRODUCTION

    A. Occurrence and Production

        Manganese is found in many, widely scattered minerals. It is about
0.1% of the earth's crust. Major commercial ores are pyrolusite (MnO_2),
manganite ($Mn_2O_3 \cdot H_2O$), hausmannite (Mn_3O_4), rhodochrosite ($MnCO_3$) and
psilomelane [$BaMn_9O_{16}(OH)_4$]. The crude metal (97-98%) is produced by
aluminum reduction of ore. Purer products are produced by electrolysis of
manganese sulfate. Other products include ferromanganese ($\geq$ 80% Mn),
spiegeleisen (a high carbon steel with up to 20% Mn), silicomanganese (65-70%
Mn, 15-20% Si, 4-6% Fe), and manganese dioxide (NAS 1973; Stokinger 1981).

    B. Uses

        The major use of manganese is in iron alloys. It is also used in
nonferrous alloys (such as manganese bronze), in dry cells (as MnO_2), and in
various chemicals, especially potassium permanganate ($KMnO_4$) and other
oxidizers (Stokinger 1981).

    C. Chemistry

        The chemistry of manganese is similar to that of iron, but is
complicated by the existence of seven oxidation states: 0, +1, +2, +3, +4,
+6, and +7. The +2 and +4 states are the most common (Stokinger 1981).

II. EXPOSURE AND EXPOSURE LIMITS

    A. Oral

        Manganese is widely distributed in foods and water. Good sources
of manganese are nuts, whole cereals, and teas. Meat, fish, and dairy products are poor sources (Crounse et al. 1983a). It is estimated that normal
consumption is 3 to 7 mg/day (NAS 1973). Manganese levels in the finished
water of 85 U.S. cities ranged from < 5.0 to 350 µg/L (USEPA 1975; cited by
NAS 1977). A study of tap water from Dallas, Texas, found an average
manganese concentration of 3.7 µg/L with a range of 1 to 8 µg/L (NAS 1977).
In 380 finished-water samples, manganese was detected in 58.7% with a mean
concentration of 25.5 µg/L and a range of 0.5 to 450 µg/L (Kopp 1970; cited
by NAS 1977).

    B. Inhalation

        Exposure is primarily occupational, and is the main route to known
manganese intoxication (NAS 1973). The threshold limit value (TLV) as a
ceiling value in workroom air is 5 mg Mn/m³ for manganese dusts and

compounds. For manganese fumes, the TLV is 1 mg/m^3 as a time-weighted average with a short-term exposure limit of 3 mg/m^3 over 15 minutes (ACGIH 1983).

Manganese releases to the ambient air are contributed by coal burning, fuel oil burning, steelmaking, and battery manufacture (Crounse et al. 1983). Snyder et al. (1975) consider airborne manganese intake, however, to be negligible.

C.   Dermal

No data.

D.   Total Body Burden and Balance Information

The total body burden is 12 to 20 mg manganese (Crounse et al. 1983). Snyder et al. (1975) estimate 12 mg for 70-kg reference man. His manganese balance is estimated as follows:

Intake (mg/day)		Losses (mg/day)	
Food and fluids	3.7	Urine	0.03
Airborne	0.002	Feces	3.6
		Sweat	0.039
		Hair, nails	0.002

III.   TOXICOKINETICS

A.   Absorption

Oral absorption of manganese in the diet is slow and incomplete: about 1 to 4% (Crounse et al. 1983). However, inhaled manganese is rapidly toxic indicating faster absorption from the lung (Snyder et al. 1975).

B.   Distribution

1.   General

Manganese is widely distributed among the organs (NAS 1973; Stokinger 1981). Highest manganese concentrations are in the bone, liver, and kidney and within cells, concentrations are higher in the mitochondria than in the cystosol and other organelles (Crounse et al. 1983).

2.   Blood

Manganese is transported in the blood serum by a β-globulin. Some researchers believe that the globulin is specific for manganese and call it transmanganin. Others report binding to transferrin as well. Concentrations in erythrocytes are reported to be increased in rheumatoid arthritis (Crounse et al. 1983).

Manganese concentrations are 25 times higher in the red blood cells than in the serum. Although a significant relation between manganese exposure and blood concentrations in workers was not found by two research groups, such a relation has been found in monkeys and rats inhaling MnO_2. Even so, no correlation may be possible with the severity of chronic toxicity symptoms because of the differences in individual susceptibility (Lauwerys 1983).

Manganese errors in analysis of blood serum or plasma may be as large as 15 or 20% if sampling containers are not thoroughly cleaned. The true mean value for Mn in serum determined after careful elimination of sources of contamination is 0.587 ± 0.183 µg/L (range $\sim$ 0.36 to 1.04 µg/L) although other literature mean values range from 0.97 to 34.3 µg/L. In neutron activation analysis of manganese, the reaction $^{56}Fe(n,p)^{56}Mn$ in packed blood cells introduces an $\sim$ 50% error. The true value of Mn is $\sim$ 0.015 µg Mn/g wet weight in packed blood cells (Versieck 1984).

3. Adipose

Of the 12 mg total body burden of reference man and of the 7.2 mg in soft tissues, adipose tissue is estimated to contain 0.5 mg (0.03 µg/g) (Snyder et al. 1975).

C. Excretion

Excretion of absorbed manganese is virtually entirely through the bile (NAS 1973). However, worker exposure has been roughly correlated with manganese concentrations in urine. Toxicity may occur when urinary concentration exceed 40 to 50 µg Mn/L (normal: < 3 µg/L) (Lauwerys 1983).

IV. EFFECTS

A. Acute and Other Short-Term Exposures

Human manganese toxicity has only been shown on exposure to high levels in the air (NAS 1980).

Inhalation of large doses of manganese compounds, especially the higher oxides, can be lethal; the mechanism seems to be local irritation, and the reaction to it (Stokinger 1981). Data are minimal.

Ingested manganese exhibits low toxicity in animals. Signs of toxicity are not evident until dietary concentrations exceed 0.1% (NAS 1980).

B. Chronic Exposure

Chronic manganese toxicity is well known in miners, millworkers, and others exposed to dust and fumes, as well as excessive manganese in well-water (often in mining villages). The usual signs and symptoms involve the central nervous system. Onset is insidious, with apathy, anorexia, and

asthenia. Then comes the characteristic manganese psychosis, with unaccount-
able laughter, euphoria, impulsiveness, and insomnia, followed by overpower-
ing somnolence. This may be accompanied by headache, recurring leg cramps,
sexual excitement followed by impotence, and/or other, rarer, effects. After
that, the victim develops speech disturbances, which may become incoherence
or even muteness. A masklike facies and general clumsiness of movement,
especially gait, and micrographia set in. This stage is often clinically
indistinguishable from Parkinson's diseases. Although patients may be
totally disabled, the syndrome is not lethal.

A completely separate disease, from inhalation only, is "manganese
pneumonia," which can be fatal. This is primarily lobar pneumonia, and can
rapidly progress to a very severe state. Fibrosis of the lung tissue has
also been reported (NAS 1973; Stokinger 1981).

C. Biochemistry

1. Effects on Enzymes

Manganese, primarily divalent, is widely distributed in the body.
It is part of several enzymes (such as pyruvate carboxylase) and can substitute for magnesium in many (NAS 1973). Enzyme systems in which manganese
is essential are involved in protein and energy metabolism and in mucopolysaccharide formation (NAS 1980).

2. Metabolism

No data.

3. Antagonisms and Synergisms

There are interactions with other divalent metal ions, especially
magnesium. Manganese-induced parkinsonism is clinically identical to the
idiopathic disease and can be successfully treated by the same means (NAS
1973; Stokinger 1981).

4. Physiological Requirements

Manganese is an essential trace element in animals. The recommended
daily allowance for humans is 2.5 to 5 mg (NAS 1980) although manganese deficiency has not been described for free-living humans.

D. Specific Organs and Systems

Except for the parkinsonism cited above, no data.

E. Teratogenicity

Maneb (manganese ethylenebisdithiocarbamate) was weakly teratogenic
to rats, but this was probably due to the organic moiety (Stokinger 1981).
Manganese deficiency in rats and mice leads to defective otolith formation
and subsequent ataxia in the offspring (Shepard 1980).

F.   Mutagenicity

No data.

G.   Carcinogenicity

No data.

V.   REFERENCES

ACGIH. 1983. TLVs® Threshold limit values for chemical substances and physical agents in the work environment with intended changes for 1983-84. Cincinnati, Ohio: American Conference of Governmental Industrial Hygienists.

Crounse RG, Pories WJ, Bray JT, Mauger RL. 1983. Geochemistry and man: Health and disease 1. Essential elements. In: Applied environmental geochemistry. Thorntain, I., ed., London, UL: Academic Press, pp. 309-333.

Lauwerys RR. 1983. Chapter II. Biological monitoring of exposure to inor- ganic and organometallic substances. In: Industrial chemical exposure: Guidelines for biological monitoring. Davis, CA: Biomedical Publications, pp. 9-50.

NAS. 1973. Natl. Academy of Sciences. Manganese. Washington, D.C.: Printing and Publishing Office, National Academy of Sciences.

NAS. 1977. Natl. Academy of Sciences. Drinking water and health. Washington, DC: Printing and Publishing Office, National Academy of Sciences.

NAS. 1980. Nat Academy of Sciences. Recommended dietary allowances. 9th ed. Washington, DC: Printing and Publishing Office, National Academy of Sciences.

Shepard TH. 1980. Catalog of Teratogenic Agents, 3rd ed. Baltimore: The Johns Hopkins University Press.

Stokinger HE. 1981. Chapter 29. The metals. In: Patty's industrial hygiene and toxicology, 3rd ed., volume IIA, Toxicology. Clayton GD, Clayton FE, eds. New York: Wiley-Interscience, John Wiley & Sons, pp. 1493-2060.

Versieck J. 1984. Trace element analysis - A plea for accuracy. Trace Elements Med 1(1):2-12.

MAMMALIAN TOXICITY SUMMARY

I. INTRODUCTION

A. Occurrence and Production

Mercury is found throughout the earth's crust, primarily as various sulfides. The main ore is cinnabar (α-HgS). It is produced by roasting the ore with an excess of oxygen, then distilling the metallic mercury to purify it (Berlin 1977; Stokinger 1981).

B. Uses

The major uses are in electrical equipment and in the chlor-alkali industry. Other uses include as a catalyst in polyurethane foams, industrial and control instruments, dental preparations, and detonators. Now obsolete uses include medicinal preparations, felting (for beaver hats, etc.), gold refining, and as an agricultural and industrial biocide (Tsuchiya 1977; Stokinger 1981).

C. Chemical and Physical Properties

Mercury is the only metal that is liquid at normal temperatures. It forms many mercuric (+2) compounds and some mercurous (+1) compounds; some of the latter disproportionate to Hg(0) plus Hg(II). Except for the chloride and acetate, water solubility varies from low to negligible. Organomercurials are also well-known; the mercury is usually divalent (Stokinger 1981).

II. EXPOSURE AND EXPOSURE LIMITS

Mercury circulation is a global phenomenon, occurring mainly through the atmosphere. A number of reviews are available (USEPA 1980).

A. Oral

The most studied source of mercury is from fish, where it is mainly present as methylmercury. Fish levels are usually below 200 ng/g, while nonfish foods are usually below 20 ng/g (USEPA 1980). A survey of 273 water supply systems found no mercury or less than 1.0 µg/L in 95.5% of them; eleven had concentrations of 1 to 4.8 µg/L (Hammerstrom et al. 1972; cited by NAS 1977). The national interim primary drinking water standard for mercury is 2 µg/L (USEPA 1975), which is higher than the 1980 criterion: 0.144 µg/L (USEPA 1980).

It has been estimated that total intake of inorganic mercury is less than 10 µg/day (Berlin 1977). However, see Part II.D.

B. Inhalation

Airborne mercury is predominantly elemental. The average concentration seems to be 20 ng/m^3, with variations from 0.5 to 50 ng/m^3, and some exceptional areas (such as near mercury mines) up to 0.8 mg/m^3 (Berlin 1977; USEPA 1980).

The threshold limit value as a time-weighted average in workroom air for an 8-h day is 0.1 mg/m^3 for mercury, aryl and inorganic compounds; 0.05 for mercury vapor (all forms except alkyl); and 0.01 mg/m^3 for organo-(alkyl) mercury; the short-term exposure limit for organo(alkyl)mercury is 0.03 mg/m^3/15 min (ACGIH 1983). The OSHA permissible exposure limit is 0.01 mg/m^3 for organo(alkyl)mercury (OSHA 1981).

C. Dermal

Except for occupational and obsolete medical uses (e.g., mercurochrome and merthiolate), dermal exposure is nil. Mercury compounds used to disinfect diapers may have caused mercury poisoning in infants (pink disease or acrodynia) (Dorland 1974).

D. Total Body Burden and Balance Information

According to Snyder et al. (1975), the mercury balance for a 70-kg reference man, in µg/day, is: intake 15 from food and fluids and 1 from air; losses 0 to 35 in urine, 10 in feces, and ∿ 1 by other routes. The soft tissue body burden was estimated to be 13 mg.

III. TOXICOKINETICS

A. Absorption

Elemental mercury is well absorbed by inhalation, but poorly absorbed after ingestion. Inorganic mercury compounds are absorbed after ingestion and, in part, after dermal application. Organomercurials (especially short-chain alkyls) are well absorbed by all routes (Berlin 1977; USEPA 1980).

B. Distribution

1. General

Once absorbed, mercury is generally distributed about the body, binding to the sulfhydryl groups of many proteins (Berlin 1977; USEPA 1980). Kidney and brain are major depots after Hg vapor exposure; kidney, after exposure to inorganic salts (Lauwerys 1983).

2. Blood

Snyder et al. (1975) estimated that 70-kg reference man contains 26 µg mercury in his whole blood (9 µg in plasma and 17 µg in erythrocytes. Methylmercury is found mainly (∿ 90%) in erythrocytes.

If exposure has occurred for $\geq$ 1 yr, mercury concentrations in blood, urine, and saliva correlate with recent exposure to mercury vapor. In chronic exposure, blood or urinary concentrations are correlated with signs of mercury vapor toxicity. Air exposure levels of 50 µg Hg/m^3 correspond to blood levels of $\sim$ 30 to 35 µg/L (controls generally have < 20 µg/L). When blood concentrations exceed 30 µg/L, some clinical evidence of toxicity is seen: a greater incidence of excreting high-molecular weight proteins and some enzymes in the urine (Lauwerys 1983).

Persons chronically exposed to alkylmercury exhibit paresthesia and sensory disturbances as the earliest poisoning signs. These occur at blood and hair levels > 200 µg/L and 50 µg/g, respectively. In such persons, whole blood concentrations should not exceed 100 µg/L (Lauwerys 1983).

Schuckmann (1979) examined 39 chlor-alkali workers exposed to mercury > 7 yr. Average workplace air concentrations were < 0.1 mg/m^3 when measured in the 1970s. In 1977, the mean blood concentration was 19.9 µg Hg/L (determined by flameless atomic absorption; poor description of methodology). No significant biochemical or neurological differences were seen between those workers and the controls.

In the general population, Kershaw et al. (1980) point out that for 2 to 3 days after ingestion of fish contaminated by methylmercury, misleadingly high mercury concentrations may be found in the blood. Whole blood concentrations for six individuals eating a single meal of fish containing 18 to 22 µg Hg/kg body weight rose from 1.6-7.5 µg/L to 40-60 µg/L with peak concentrations appearing 4.7 to 14.0 h after ingestion (analytical methodology good). These same researchers (Phelps et al. 1980) reported that hair was a better indicator of past mercury exposure in Indians and some Caucasians living on Canadian reservations who ate methylmercury-contaminated fish.

Turner et al. (1982) studied a Peruvian population chronically exposed to methylmercury in ocean fish. Methylmercury concentrations were 11 to 275 µg/L blood (mean 82 µg/L) for this group compared to 3.3 to 25.1 µg/L (mean 9.9 µg/L) for a control population who ate $\sim$ 20% as much fish. The incidence of parathesias was higher in the control group (49.5% vs. 29.5%)! Symptoms or signs attributable to methylmercury poisoning were not identified in any individual.

Sherlock et al. (1984) found a nearly linear relation between methylmercury intake (40 to 230 µg/day) and blood Hg concentration. At equilibrium, for each 1 µg methylmercury ingested per day, the blood Hg concentration increases 0.8 µg/kg.

3. Adipose

Snyder et al. 1975) estimated 4.5 mg Hg (0.30 µg/g) in adipose tissue of reference man.

C. Excretion

Mercury is excreted in the urine and feces. Small quantities go into the hair and other routes, including the exhaling of some elemental mercury (Berlin 1977; USEPA 1980).

IV. EFFECTS

A. Acute and Other Short-Term Exposures

Acute mercury poisoning is usually by the soluble inorganic salts. Early signs and symptoms include pharyngitis, dysphagia, abdominal pain, nausea and vomiting, bloody diarrhea, and shock. Later swelling of the salivary glands, stomatitis, loosening of the teeth, nephritis, anuria, and hepatitis occur. Death results from the effects on the gastrointestinal tract (ulcerations, bleeding, shock) and/or kidney (Berlin 1977; Stokinger 1981).

High levels of mercury vapor are extremely irritating to the lung and cause erosive bronchitis and bronchiolitis with interstitial pneumonia (Berlin 1977; Gerstner and Huff 1977).

B. Chronic Exposure

Mercurialism, chronic intoxication by elemental mercury vapor or mercury salts, is much more frequently seen than acute toxicity, due to its cumulative nature. The classic syndrome begins with psychic and emotional disturbances: the victim becomes excitable and irascible, especially when criticized. He can no longer concentrate mentally, and becomes depressed. He may complain of headache, fatigue, weakness, loss of memory, drowsiness, or insomnia. He will show a fine muscular tremor, usually beginning in the hands, which soon interferes with handwriting and other precision work. These are "intention" tremors, absent during sleep, most pronounced when under stress. The victim may develop paresthesias and neuralgia. Other systems are sometimes affected; renal disease, chronic nasal catarrh, and epitaxis are common. Gingivitis, stomatitis (sometimes severe), digestive disturbances, and ocular lesions are seen (Gerstner and Huff 1977; Stokinger 1981).

With mercury salts, the picture is similar, but kidney toxicity is more important, frequently being the cause of death.

Methylmercury (and other alkylmercury compounds) produce "Minamata Disease," which has the clinical appearance of encephalitis. The earliest signs are gradual decreases in the senses of touch, vision, hearing, and taste. The skin becomes less and less sensitive; numbness in the fingers, toes, lips, and tongue interferes with normal activities, including walking, working, speaking, eating, and drinking. Tunnel vision occurs and may lead to complete blindness. Hearing loss is over the entire frequency range. The motor system is affected by progressive loss of balance, tremors, and incoordination. Mood changes like those observed in mercurialism occur. Prenatal intoxication occurs; symptoms are often more obvious in the child than in the mother (Gerstner and Huff 1977; USEPA 1980).

C. Biochemistry

1. Effects on Enzymes

Mercury apparently acts by reaction with sulfhydryl groups in proteins and, possibly, smaller molecules. The details are not understood, but mercury salts are almost general-purpose enzyme inhibitors (USEPA 1980; Berlin 1977).

2. Metabolism

Metallic mercury and organomercurials are transformed to divalent mercury (Berlin 1977; USEPA 1980). Short-chain alkyl compounds are resistant to biotransformation, but aryl and aryloxyalkyl derivatives liberate inorganic mercury (Lauwerys 1983).

3. Antagonisms and Synergisms

Mercury intoxication is treated by chelation, preferably with N-acetyl-dℓ-penicillamine (Gerstner and Huff 1977; Stokinger 1981).

D. Specific Organs and Systems

As described above, the main organs affected by mercury are the brain and kidney. In both, mercury produces destruction of ultrastructural elements and, eventually, degeneration of cells and tissues. Metallic mercury and organomercurials affect the brain more, especially the cerebellum and some parts of the cerebral cortex. Inorganic mercury salts affect the kidney more, especially the tubules (Berlin 1977; USEPA 1980).

E. Teratogenicity

Despite its prenatal toxicity, methylmercury is not teratogenic to humans. Limited animal studies report that methylmercury and a mercury salt were teratogenic (USEPA 1980).

F. Mutagenicity

Methylmercury may cause chromosome aberrations; no other studies have been reported (USEPA 1980).

G. Carcinogencity

There is no evidence.

V. REFERENCES

ACGIH. 1983. TLVs® threshold limit values for chemical substances and physical agents in the work environment with intended changes for 1983-84. Cincinnati, Ohio: American Conference of Governmental Industrial Hygienists.

Berlin M. 1977. Mercury. In: Toxicology of metals - Vol. II. Springfield, VA: National Technical Information Service, pp. 301-344, PB 268-324.

Dorland. 1974. Dorland's illustrated medical dictionary, 25th ed. Philadelphia, PA: W. B. Saunders.

Gerstner HB, Huff JE. 1977. Clinical toxicology of mercury. J Toxicol Environ Health 2:491-526.

Kershaw TG, Dhahir PH, Clarkson TW. 1980. The relationship between blood levels and dose of methylmercury in man. Arch Environ Health 35(1):28-36.

Lauwerys RR. 1983. Chapter II. Biological monitoring of exposure to inorganic and organometallic substances. In: Industrial chemical exposure: guidelines for biological monitoring. Davis, CA: Biomedical Publications, pp. 9-50.

NAS. 1977. National Academy of Sciences. Drinking water and health. Vol. I. Washington, DC.

NIOSH. 1973. Natl. Inst. Occupational Safety and Health. Criteria for a recommended standard: occupational exposure to mercury. Washington, DC: U.S. Government Printing Office, Publication No. HSM 73-11024.

OSHA. 1981. Occupational Safety and Health Admin. Occupational safety and health standards. Subpart 2--Toxic and hazardous substances. Code of federal regulations 29 (Part 1910.1000) pp. 673-679.

Phelps RW, Clarkson TW, Kershaw TG, Wheatley B. 1980. Interrelationships of blood and hair mercury concentrations in a North American population exposed to methylmercury. Arch Environ Health 35(3):161-168.

Schuckmann F. 1979. Study of preclinical changes in workers exposed to inorganic mercury in chloralkali plants. Int Arch Occup Environ Health 44(3):193-200.

Shepard TH. 1980. Catalog of teratogenic agents, 3rd ed. Baltimore: The Johns Hopkins University Press.

Sherlock J, Hislop J, Newton D, Topping G, Whittle K. 1984. Elevation of mercury in human blood from controlled chronic ingestion of methylmercury in fish. Hum Toxicol 3(2):117-131.

Snyder WS, Cook MJ, Nasset ES, Karhausen LR, Howells GP, Tipton IH. 1975.
International Commission on Radiological Protection. Report of the task
group on reference man. New York. ICRP Publication 23.

Stokinger HE. 1981. Chapter 29. The metals. In: Patty's industrial
hygiene and toxicology. 3rd ed. Volume 2A. Clayton GD, Clayton FE, eds.
New York: A Wiley-Interscience Publication. John Wiley and Sons,
pp. 1493-2060.

Turner MD, Marsh DO, Smith JC et al. 1982. Methylmercury in populations
eating large quantities of marine fish. Arch Environ Health 35(6):367-378.

USEPA. 1975. U.S. Environmental Protection Agency. National interim
primary drinking water regulations. (40 FR 59566).

USEPA. 1980. U.S. Environmental Protection Agency. Ambient water quality
criteria for mercury. Springfield, VA: National Technical Information
Service, PB 81-117699.

MOLYBDENUM

MAMMALIAN TOXICITY SUMMARY

I. INTRODUCTION

A. Occurrence and Production

Molybdenum minerals are found in many areas. Commerical ores include molybdenite (MoS_2), powellite ($CaMoO_4$), and wulfenite ($PbMoO_4$). Much comes as a by-product of copper production. Ore concentrates are roasted to produce the oxide (MoO_3), which is reduced to the metal with hydrogen (Stokinger 1981).

B. Uses

Most molybdenum is used as the oxide to alloy steel. Other uses include as a lubricant (MoS_2), as catalysts, as pigments (especially molybdates), and in ceramics (Stokinger 1981).

C. Chemical and Physical Properties

Molybdenum has a very complicated chemistry, including valences from 0 to +6 (the last is usually the most stable), nine oxides (white, red, violet, and black colors), multiple coordination numbers (4, 6, 8) and other peculiarities (Stokinger 1981).

II. EXPOSURE AND EXPOSURE LIMITS

A. Oral

Daily intake has been estimated at 100 to 500 µg. Highest food levels are in leafy vegetables and legumes (Friberg 1977). The main dietary contributors of molybdenum (based on amounts actually consumed) are meat, grains, and legumes (NAS 1980).

Molybdenum was found in finished water of the 100 largest U.S. cities at a median concentration of 1.4 µg/L, with a range from not detectable to 68 µg/L (Durfor and Becker 1964; cited by NAS 1977). In a study of 380 finished waters, 29.9% had detectable molybdenum levels ranging from 3 to 1,024 µg/L with a mean of 85.9 µg/L (Kopp 1970; cited by NAS 1977).

B. Inhalation

The threshold limit value (TLV) as a time-weighted average in workroom air for an 8-h day is 10 mg Mo/m^3 for insoluble molybdenum; the short-term exposure limit (STEL) is 20 mg/m^3 for 15 min. The TLV for soluble molybdenum compounds is 5 mg/m^3; the STEL is 10 mg/m^3 for 15 min (ACGIH 1983). The OSHA permissible exposure limit is 15 mg/m^3 for insoluble molybdenum compounds and 5 mg/m^3 for soluble compounds (OSHA 1981).

C. Dermal

Even occupationally exposed persons do not exhibit dermatoses from contact with the less soluble, less toxic forms of molybdenum (Stokinger 1981).

D. Total Body Burden and Balance Information

According to Snyder et al. (1975), the molybdenum balance for 70-kg reference man in µg/day is: intake 300 from food and fluids and < 0.1 from air; losses 150 in urine, 120 in feces, and 20 by other routes.

III. TOXICOKINETICS

A. Absorption

Molybdenum, in limited studies, is well absorbed from the gastrointestinal tract, but not by inhalation (Friberg 1977).

B. Distribution

1. General

Molybdenum is generally distributed, with greatest concentrations in the kidney, liver, bones, and pancreas (Friberg 1977). Tissue levels are proportional to the dietary sulfate intake (Stokinger 1981).

2. Blood

In humans injected with ^{99}Mo, the radioisotope cleared the blood rapidly. At 24 h, < 0.5% was still in the blood. Whole blood concentrations were greater than plasma concentrations. Molybdenum is firmly bound to the erythrocytes and the plasma proteins. In anemia, Mo concentrations decrease in both red cells and plasma. Normal mean blood values for Mo (detected in 48 of 229 samples) were 0.50 to 15.73 µg/100 mL. High values were found in samples from Missoula, Montana (41 µg/100 mL) and Rapid City, South Dakota (14.2 µg/L) (Stokinger 1981). Molybdenum determinations in blood serum are subject to gross errors due to sample contamination according to Versieck (1984).

3. Adipose

In dogs injected with ^{99}Mo, negligible amounts of molybdenum were found in fat (Stokinger 1981).

C. Excretion

Excretion of absorbed molybdenum is primarily in the urine, with some in the bile, sweat, etc. (Friberg 1977).

IV. EFFECTS

A. Acute and Other Short-Term Exposures

Acute toxicity is negligible for MoS_2 and for the molybdates. Hexavalent compounds such as MoO_3 are lethal at oral doses of 100 mg/day. Rabbits fed $\geq 0.1\%$ $Na_2MoO_4 \cdot 2H_2O$ in the diet died within a few weeks. MoO_3 dust at 164 mg Mo/m^3 for 1 h/day was extremely irritating to guinea pigs; half had died by the 10th exposure. Signs included loss of appetite, weight, and hair, diarrhea, and muscular incoordination. Livers and kidneys of severely poisoned animals undergo fatty degeneration (Stokinger 1981).

B. Chronic Exposure

Chronic toxicity is seen as loss of appetite, listlessness, diarrhea, and reduced growth rate. Anemia is common, and animals have deformities of joints and long bones as well as mandibular exostoses. Cattle on diets high in molybdenum and low in copper develop "teart disease," which also produces impaired reproduction. A gout-like disease (perhaps from increased xanthine oxidase activity) has been seen in people living in a high-Mo area of Armenia. Dietary intakes there are 10 to 15 mg/day. One survey of smelter workers found elevated serum ceruloplasmin and serum uric acid. Dietary levels of 0.54 mg Mo/day have been associated with increased copper excretion in the urine (Friberg 1977; Stokinger 1981; NAS 1980).

C. Biochemistry

1. Effects on Enzymes

Molybdenum is a constituent of three metalloflavoproteins: xanthine oxidase, aldehyde oxidase, and sulfite oxidase (Friberg 1977).

MoO_4^{2-} inactivates glutaminase and sulfoxidase (Stokinger 1981).

2. Metabolism

No information.

3. Antagonisms and Synergisms

The antagonism between copper and molybdenum is affected by sulfate in the diet depending on the species. For animals with normal copper stores, sulfate prevents molybdenum toxicity. Thiol compounds reduce molybdenum toxicity. Isomorphic WO_4^{2-} antagonizes MoO_4^{2-} in the chick and the rat. Molybdenum acts synergistically with fluoride to decrease caries incidence in rats (Stokinger 1981).

4. Physiological Requirements

Molybdenum is an essential trace mineral in ruminants and plants. Deficiencies are unknown in humans so the diet must supply sufficient molybdenum to carry out its roles in enzyme functions. The recommended daily allowance estimated to be safe and adequate is 0.15 to 0.5 mg (NAS 1980).

D. Specific Organs and Systems

   1. Blood

      Toxic doses of molybdenum cause anemia and derange copper
metabolism.

   2. Liver and Kidney

      Fatty infiltration of kidneys and liver occurs in poisoned animals.

E. Teratogenicity

      No defects were found in chick embryos injected with lethal doses
of molybdenum. Some runting occurred in third-generation offspring after
mice were fed a diet containing 0.45 ppm Mo (Shepard 1980).

F. Mutagenicity

      No data.

G. Carcinogencity

      No data.

V. REFERENCES

ACGIH. 1983. TLVs® Threshold limit values for chemical substances and
physical agents in the work environment with intended changes for 1983-84.
Cincinnati, Ohio: American Conference of Governmental Industrial Hygienists.

Friberg L. 1977. Molybdenum. In: Toxicology of metals - Vol. II.
Springfield, VA: National Technical Information Service, pp. 345-357,
PB 268-324.

NAS. 1977. National Academy of Sciences. Drinking water and health.
Vol. I. Washington, DC.

NAS. 1980. National Academy of Sciences. Recommended dietary allowances.
9th ed. Washington, DC: Printing and Publishing Office, National Academy of
Sciences.

OSHA. 1981. Occupational Safety and Health Admin. Occupational safety and
health standards. Subpart 2--Toxic and hazardous substances. Code of
federal regulations 29 (Part 1910.1000) pp. 673-679.

Shepard TH. 1980. Catalog of teratogenic agents, 3rd ed. Baltimore: The
Johns Hopkins University Press.

Snyder WS, Cook MJ, Nasset ES, Karhausen LR, Howells GP, Tipton IH.   1975.
International Commission on Radiological Protection.   Report of the task
group on reference man.   New York.   ICRP Publication 23.

Stokinger HE.   1981.   Chapter 29.   The metals.   In: Patty's industrial
hygiene and toxicology.   3rd ed.   Volume 2A.   Clayton GD, Clayton FE, eds.
New York: A Wiley-Interscience Publication.   John Wiley and Sons,
pp. 1493-2060.

NICKEL

MAMMALIAN TOXICITY SUMMARY

I. INTRODUCTION

A. Occurrence and Production

Nickel constitutes about 0.008% of the earth's crust, predominantly in igneous rocks. The earth's core contains 8.5% nickel. A variety of ores are of importance: the sulfide ores pentlandite, $(Fe,Ni)_9S_8$; chalcopyrite, $CuFeS_2$, and pyrrhotite, Fe_xS_x+1; the arsenides $NiAs$, $NiAs_2$, and $NiAsS$; and the antimonides $NiSb$ and $NiSbS$. Many ores also contain other sought metals, including Co, Cu, Au, Ag, and the Pt-group metals. The smelting and refining process varies with the other constituents, but often involves roasting to the oxide, NiO, and conversion to the volatile nickel carbonyl, $Ni(CO)_4$, which is then reduced to the pure metal. Other processes include electrolytic refining (NAS 1975; Stokinger 1981).

B. Uses

About half of all nickel is used in steels, mostly in the "stainless steels." A fourth is used in other alloys for use in many applications requiring corrosion and temperature resistance. Some is electroplated to provide a tarnish-resistant surface and also as a strike coat before precious-metal plating. Other uses include coinage (replacing silver), as a catalyst, in ceramics, in some pigments, and in Ni-Cd batteries (Stokinger 1981).

C. Chemistry

Most nickel compounds have +2 valence, but +1, +3, and +4 are also known. The common salts are reasonably water-soluble (EPA 1980; Stokinger 1981).

II. EXPOSURE AND EXPOSURE LIMITS

A. Oral

The major source of uptake by the average person is in food. This varies up to 900 µg/day, but is typically 300 to 500 µg/day (NAS 1975). A study of 969 U.S. water supplies found the average nickel level to be 4.8 µg/L (McCabe undated; cited by NAS 1975). Tap water from Hartford, Connecticut, had a mean concentration of nickel of 1.1 µg/L compared to 200 µg/L for Sudbury, Ontario, a major nickel-producing area (McNealey et al. 1972; cited by NAS 1977). USEPA (1980) recommends a drinking water criterion of 13.4 µg Ni/L.

B.   Inhalation

Airborne nickel seems to come primarily from combustion of coal and petroleum products. Recently levels have been decreasing, with levels averaging about 9 ng/m³ in urban areas and 2 ng/m³ in nonurban areas (USEPA 1980). The threshold limit value (TLV) as a time-weighted average in workroom air for an 8-h day is 0.1 mg/m³, as nickel, for soluble nickel compounds; the short-term exposure limit is 0.3 mg/m³ for 15 min. The TLV for nickel metal is 1 mg/m³ (ACGIH 1983). The OSHA permissible exposure limit (PEL) is 1 mg Ni/m³ (OSHA 1981). However, NIOSH (1977) has recommended 0.015 mg Ni/m³ as the limit. The TLV for Ni(CO)₄ is 0.35 mg/m³ (ACGIH 1983) and its OSHA PEL is 0.007 mg/m³ (OSHA 1981). Ni(CO)₄ is not stable enough to have any potential for injury to public health.

C.   Dermal

Dermal absorption occurs, but is significant only to allergenicity, discussed below (NAS 1975; EPA 1980). Contact with nickel-plated jewelry and other items is a frequent cause of contact dermatitis in the general public.

D.   Total Body Burden and Balance Information

According to Snyder et al. (1975), the nickel balance for a 70-kg reference man in µg/day is: intake 400 from food and fluids and 0.6 from air; losses 11 in urine, 370 in feces, and 21 by other routes. Total body burden is about 10 mg, with wide variations (NAS 1975). About 5.3 mg Ni is the soft tissue burden (Snyder et al. 1975).

III. TOXICOKINETICS

A.   Absorption

The degree of absorption of ingested nickel generally depends on the solubility of the compound (typically 1 to 10% of the ingested dose). The rate of absorption of lung-deposited particles is also generally dependent on water solubility (NAS 1975; USEPA 1980).

B.   Distribution

1.   General

Relatively high concentrations are found in the kidney, liver, and brain (NAS 1975). Human tissues contain ~ 0.02 to 1.5 ppm Ni (Crounse et al. 1983).

2.   Blood

Snyder et al. (1975) estimate 0.16 mg nickel in whole blood (0.09 mg in plasma and 0.07 mg in red blood cells) of 70-kg reference man. Absorbed nickel is transported by a plasma protein called "nickeloplasmin."

Unexposed individuals have serum nickel levels about 2 to 3 µg/L; those exposed to toxic doses have higher dose-related levels (NAS 1975).

Normal blood plasma nickel concentrations are < 10 µg Ni/L (usually 2 to 4). A level of 10 µg Ni/L has been proposed as the critical concentration for workers (Lauwerys 1983). Literature reports of normal nickel concentrations in blood serum are often grossly in error due to sample contamination according to Versieck (1984).

At a nickel refinery in Norway, workers engaged in electrolysis, who were exposed to soluble nickel compounds, had an av. 7.4 µg Ni/L in their blood plasma; roasting and smelting department workers had an av. 6 µg Ni/L. University students serving as controls had 4.2 µg Ni/L in their plasma. Usually, urinary nickel does not correlate well with degree of exposure or health status although it has been used as an index of exposure. Blood monitoring was not recommended by NIOSH for medical surveillance of workers exposed to inorganic nickel (NIOSH 1977). Hogetveit et al. (1980)* concluded that average nickel in plasma in Norwegian nickel refinery workers correlates better with degree of exposure than nickel in urine unless a 24-h urine sample is collected. Two plasma measurements are averaged: morning and after-work.

3.   Adipose

Snyder et al. (1975) estimate 0.52 mg Ni (0.035 µg/g) in adipose tissue of reference man.

C.   Excretion

The main excretion route for absorbed nickel is urine with normal values of 2 to 4 µg/L. Some nickel also appears in sweat and hair (USEPA 1980).

IV.   EFFECTS

A.   Acute and Other Short-Term Exposures

The major acute toxicity of nickel, both the metal and compounds, are dermatoses: contact dermatitis, atopic dermatitis, and allergic sen-sensitization. Except for inhalation of nickel carbonyl (which is highly irritating to pulmonary tissue, sometimes causing death from pulmonary edema), nickel and its compounds have little toxicity (NAS 1975; USEPA 1980; Stokinger 1981).

---

* Details of the analytical methodology (atomic absorption spectrophotometry) were not described in this publication but were referred to an earlier paper.

"Nickel itch" or contact dermatitis, is variable in nature, but usually begins with a sensation of burning and itching, followed by erythema and a nodular eruption, which may progress to postules or ulcers. Recovery occurs after a week or so. The syndrome is most commonly seen in the general population, especially women, with characteristic nickel sources being garter clips (before the panty hose era) and costume jewelry, especially earrings (NAS 1975; Stokinger 1981).

Some patients presenting with contact dermatitis having atopic dermatitis--a chronic pruritic eruption that continues after the putative cause (such as nickel) is removed. The relationship between the two dermatoses is unclear (NAS 1975; USEPA 1980).

Finally, nickel is capable of causing allergic sensitization. The existence of this phenomenon is evidence for some absorption of nickel through skin (NAS 1975; USEPA 1980).

B.   Chronic Exposure

The only effect of chronic exposure to nickel, which has been studied is carcinogenesis. Epidemiological studies of nickel smelter and refinery workers have shown that some compound or combination of compounds causes cancers in the respiratory tract--in the nasal cavity, and in the lung (IARC 1976; USEPA 1980; Stokinger 1981). Squamous cell carcinomas are the most common type seen at both sites. There is little evidence that other nickel workers have an increased incidence of respiratory cancers. Inhalation or parenteral administration of Ni dust, Ni subsulfide, NiO, Ni(CO)$_4$, and Ni biscyclopentadiene was carcinogenic in animals (NAS 1977).

In epidemiological studies, nickel in drinking water correlated poorly with mortality from oral or intestinal cancer and there was no correlation with respiratory cancer. NAS concluded there was no pressing need for a standard for nickel in drinking water (NAS 1977).

C.   Biochemistry

1.   Effects on Enzymes

Nickel has a variety of effects on enzymes, apparently because it can substitute, more or less ably, for other divalent metal ions, especially zinc. It is capable of both activating and inhibiting the same enzyme, depending on its concentration (NAS 1975).

2.   Metabolism

There is no evidence that nickel is metabolized by the body.

3.   Antagonisms and Synergisms

Nickel has been shown to be synergistic with certain carcinogens--polynuclear aromatic hydrocarbons and (possibly) asbestos (USEPA 1980).

4. Physiological Requirements

Nickel is an essential trace element in chicks, rats, and swine (Crounse et al. 1983).

D. Specific Organs and Systems

Nickel has most of its effects on the skin and respiratory system. These have been discussed above (Sections IV.A. and B.).

E. Teratogenicity

Nickel can cross the placental barrier, but there is no solid evidence of teratogenicity (USEPA 1980). Nickel carbonyl was teratogenic in rats; i.v. doses of $NiCl_2$ (1 to 6.9 mg/kg) on single days 7 through 11 of gestation was teratogenic in mice (Shepard 1980).

F. Mutagenicity

No data are reported.

G. Carcinogenicity

Nickel in some forms is likely to be carcinogenic to man. Excess risk of nasal cavity and lung cancers has been shown conclusively in nickel refinery workers (IARC 1976).

V. REFERENCES

ACGIH. 1983. TLVs® Threshold limit values for chemical substances and physical agents in the work environment with intended changes for 1983-84. Cincinnati, Ohio: American Conference of Governmental Industrial Hygienists.

Crounse RG, Pories WJ, Bray JT, Mauger RL. 1983. Geochemistry and man: health and disease 2. Elements possibly essential, those toxic and others. In: Applied environmental geochemistry. Thornton I., Ed. London, UK: Academic Press, pp. 309-333.

Hogetveit AC, Barton RT, Andersen I. 1980. Variations of nickel in plasma and urine during the work period. J Occup Med 22(9):597-600.

IARC. 1976. Internatl. Agency for Research on Cancer, IARC Monogr Evaluation Carcinog Risk Chem Man. 11 (Nickel and nickel compounds):75-112.

Lauwerys RR. 1983. Chapter II. Biological monitoring of exposure to inorganic and organometallic substances. In: Industrial chemical exposure: Guidelines for biological monitoring. Davis, CA: Biomedical Publications, pp. 9-50.

NAS. 1975. Natl. Academy of Sciences. Medical and biological effects of environmental pollutants: Nickel. Washington, DC: Natl. Acad. Sci.

NAS.   1977.   Natl. Academy of Sciences.   Drinking water and health.   Vol. I.
Washington, DC.

NAS.   1980.   Natl. Academy of Sciences.   Recommended dietary allowances.   9th
ed.   Washington, DC: Printing and Publishing Office, National Academy of
Sciences.

NIOSH.   1977.   Natl. Inst. Occupational Safety and Health.   Criteria for a
recommended   standard: occupational   exposure   to   inorganic   nickel.
Washington, DC: US Government Printing Office, DHEW (NIOSH) Publication No.
77-164.

OSHA.   1981.   Occupational Safety and Health Admin.   Occupational safety and
health   standards.   Subpart 2--Toxic and hazardous   substances.   Code of
federal regulations 29 (Part 1910.1000) pp. 673679.

Shepard TH.   1980.   Catalog of teratogenic agents, 3rd ed.   Baltimore: The
Johns Hopkins University Press.

Snyder WS, Cook MJ, Nasset ES, Karhausen LR, Howells GP, Tipton IH.   1975.
International Commission on Radiological Protection.   Report of the task
group on reference man.   New York.   ICRP Publication 23.

Stokinger HE.   1981.   Chapter 29.   The metals.   In: Patty's industrial
hygiene and toxicology, 3rd ed., volume IIA, Toxicology.   Clayton GD, Clayton
FE, eds.   New York: Wiley-Interscience, John Wiley & Sons, pp. 1493-2060.

USEPA.   1980.   U.S. Environmental Protection Agency.   Ambient water quality
criteria for nickel.   Springfield, VA: National Technical Information
Service.   Publ. No. PB81-117715.

Versieck J.   1984.   Trace element analysis - A plea for accuracy.   Trace
Elements Med 1(1):2-12.

NIOBIUM (COLUMBIUM)

MAMMALIAN TOXICITY SUMMARY

I. INTRODUCTION

A. Occurrence and Production

About as abundant in the earth's crust as nickel (24 ppm), niobium occurs with tantalum in the minerals tantalite, columbite, tantalocolumbite, and pyrochlore (Merck Index 1983; Stokinger 1981). Little Nb-bearing ore is mined in the United States; most is imported from Brazil, Nigeria, and Malaysia. Recovery is complex and varied in trying to separate the very similar Nb and Ta. Stokinger (1981) summarizes some of the processing.

B. Uses

Niobium or, as it is better known in the metallurgical industry, columbium is used in high-temperature steel alloys, in chrome-nickel-steel alloys to prevent the formation of chromium carbide, in some high-iron aluminum alloys, and in some permanent magnet alloys (Browning 1969). Merck Index (1983) lists these uses: ferroniobium is used in stainless steels and welding rods; high-temperature and nuclear reactions use Nb-base alloys; Nb is also used as a getter in electronic vacuum tubes.

C. Chemical and Physical Properties

Niobium, atomic no. 41, melts at 2468°C and boils at 4927°C. Niobium exhibits valences of 2, 3, 4, and most commonly, +5. $NbCl_5$ decomposes in moist air, evolving HCl. Nb_2O_5 is water insoluble (Merck Index 1983). The chemistries of tantalum and niobium are very similar (Browning 1969).

II. EXPOSURE AND EXPOSURE LIMITS

A. Oral

Schroeder and Balassa (1965; cited by Snyder et al. 1975) found 600 μg Nb/day in an institutional diet and ∿ 20 μg Nb/day in drinking water. They found niobium in almost every food analyzed: ∿ 1 μg Nb/g in cereals, meat, and dairy food and ∿ 0.7 μg Nb/g in vegetables, fruit, and fish. Higher-than-average concentrations were found in tea, coffee, pepper, and fats (not especially unusual since the water content of other materials is much higher).

B. Inhalation

No information in secondary sources except for occupational exposure in processing the ores and in forging, fabrication, and welding (Stokinger 1981). There are no occupational exposure limits.

C. Dermal

No information.

D. Total Body Burden and Balance Information

According to Snyder et al. (1975), the niobium balance for 70-kg reference man is as follows:

Intake, µg/day		Losses, µg/day	
Food and fluids	620	Urine	360
		Feces	260
		Sweat	Trace
		Hair	0.3

They speculate that the niobium burden of the soft tissue of reference man might be 110 µg.

III. TOXICOKINETICS

A. Absorption

Much information about the behavior of niobium in the body is based on studies of the radioisotope pair ^{95}Zr - ^{95}Nb,* common nuclear fission products. Their fate is usually assumed to be the same. The range of gastrointestinal absorption of the pair of radioisotopes has been reported from nil to $\sim$ 2% (Smith and Carson 1978). However, Schroeder and Balassa (1965; cited by Snyder et al. 1975) found 360 µg/day in urine and 620 µg/day intake from diet and drinking water, which would indicate that absorption is > 50% from the gastrointestinal tract.

B. Distribution

1. General

Large colloidal particles of ^{95}Zr and ^{95}Nb are compounds rapidly localized by the liver and spleen of rabbits, cats, rats, and mice. ^{95}Zr-^{95}Nb nitrate buffered by citrate 2 min after i.v. injection was rapidly distributed to lungs, liver, kidneys, muscle, skeleton, and spleen of rats. The maximum observed in the skeleton was 26% of the injected dose at 30 days. ^{95}Nb alone accumulated less in bone of mice than did ^{95}Zr-^{95}Nb. ^{95}Zr-^{95}Nb exhibit transplacental transfer in pregnant rodents (Smith and Carson 1978). After injection of ^{95}Nb, laboratory animals showed highest Nb concentrations in bone, kidney, spleen, testes, and liver (Stokinger 1981).

After inhalation, the lung-clearance rate is slow. The biologic half-life for tracer ^{95}Nb was estimated to be 120 days; in humans, the half-life for pulmonary clearance of ZrO_2 (^{95}Nb) was estimated at 224 days (Stokinger 1981).

---

* ^{95}Nb is the daughter of ^{95}Zr (Stokinger 1981).

2. Blood

Intravenous injection of colloidal compounds of ^{95}Zr-^{95}Nb are rapidly cleared from the blood within 1 min. ^{95}Zr-^{95}Nb complexes are also cleared rapidly from the blood. However, after i.p. injection in rats of carrier-free ^{95}Zr-^{95}Nb oxalates, ^{95}Nb was preferentially accumulated in the blood at 4 days whereas ^{95}Zr was preferentially accumulated in the bone (Smith and Carson 1978). Limited human analyses by Schroeder and Balassa (1965; cited by Stokinger 1981) found 0.53 to 0.74 ppm Nb in serum and 4.19 to 6.4 ppm in red cells. Human kidney, liver, aorta, lung, pancreas, testes, spleen, brain, and hair usually contained part-per-million concentrations.

3. Adipose

In their summary of numerous ^{95}Zr-^{95}Nb distribution studies, Smith and Carson (1978) found that adipose tissue was not monitored as a site for accumulation in laboratory mammals.

C. Excretion

Rats intubated with ^{95}Zr-^{95}Nb excreted the radioisotopes predominantly in the feces for the first 6 days; from day 7 to day 63, the amounts were comparable in urine and in feces. By day 63, 3.76% had been excreted in urine and 64.1%, in the feces (Smith and Carson 1978). Mice, rats, monkeys, and dogs excreted 97 to > 99% of oral doses of ^{95}Nb given alone (Stokinger 1981).

IV. EFFECTS

A. Acute and Other Short-Term Exposures

No toxic effects in humans have been reported (Browning 1969). The i.p. LD_{50} of $NbCl_5$ in rats was 14 ng Nb/kg, but K niobate was well tolerated orally. The oral LD_{50} was 1,140 mg Nb/kg (Cochran et al. 1950). The i.p. LD_{50} for male CF mice was 61 mg/kg for $NbCl_5$; the oral LD_{50}, 940 mg/kg. In mice, the i.p. LD_{50} (3 to 5 days) for K niobate was 13 mg Nb/kg; for the rat, the LD_{50} was 92 mg Nb/kg. The oral LD_{50}, however, was 725 mg Nb/kg for the rat, given as the niobate in a citrate solution, which may account for the difference from the value reported by Cochran et al. (1950) (Stokinger 1981).

Toxic symptoms of lethal doses include urination, defecation, abdominal stretching with cavitation of the lower abdominal area, a milky anal exudate, decreased respiration, and lethargy. $NbCl_5$ was severely irritating to unabraded rabbit skin but only slightly irritating to the rabbit eye (Stokinger 1981).

In the cat Nb^{5+} produced transient hypotension and ECG changes at doses > 1.0 mg/kg, with a dose of 5 mg/kg immediately giving complete cardio-vascular collapse and respiratory paralysis. Atropine and epinephrine could not counteract the effects (Stokinger 1981).

Daily repeated i.p. doses of NbCl$_5$ or K niobate in rats, rabbits, and dogs (which were eventually lethal) caused moderate weight loss and kidney changes including tubular necrosis. However, rats could eat diets containing up to 1% of either compound for 7 wk without adverse effects (Stokinger 1981).

B. Chronic Exposure

Browning (1969) stated that Schubert reported in 1947 a severe chronic intoxication in animals from 50 mg Nb/kg with no details or symptoms The 1947 report (Science 105:389-390), however, does not mention niobium (only Zr, La, Ce, and Y). Schubert (1949) himself cited the 1947 Science article as containing information on the rat toxicity of Cb: "Fifty milligrams of Cb per kilogram in the form of sodium columbate gave rise to symptoms of severe intoxication of a chronic nature." To judge from other experiments by Schubert and coworkers cited by Smith and Carson (1978), the dosing was not chronic.

Rats consuming a diet with 1.62 ppm Nb and drinking water with 5 ppm Nb as Na niobate for life developed glycosuria, reduced protein in urine, reduced longevity in males, decreased heart weight in males, and increased body weight in males. Signs developed as early as day 30. A similar lifetime study in mice showed reduced weight in older animals, shortened female lifespans, and hepatic fatty lesions. Liver toxicity, but no other effects, was observed in rats fed 0.01, 0.1, or 1% Nb in the diet after 12 wk. The liver changes included perinuclear vacuolization of the parenchymal cells and coarse granulation of the cytoplasm (Stokinger 1981).

C. Biochemistry

1. Effects on Enzymes

Nb inhibits adenosine triphosphatase (Cochran et al. 1950). An inhibitory effect on certain enzymes, e.g., succinic dehydrogenase, may be the reason niobium is more toxic to animals than tantalum and some of the rare earths (Browning 1969).

2. Metabolism

^{95}Zr-^{95}Nb are bound to the convoluted tubules in the kidney. Long-term residues in the spleen are located in the germinal lymphoid corpuscles. In the liver, spleen, and kidney, the particles to which they are bound have sedimentation numbers between 2,000 and 10,000 Svedbergs (Smith and Carson 1978).

3. Antagonisms and Synergisms

No information.

D.   Specific Organs and Systems

   1.   Skin

   $NbCl_5$ is corrosive to the skin but not the eye (Stokinger 1981).

   2.   Liver

   Fatty lesions, vacuolization of parenchymal cells, and coarse granulation of the cytoplasm have been observed in chronic dosing experiments (Stokinger 1981).

   3.   Kidney

   Tubular necrosis and other changes are observed with lethal doses (Stokinger 1981).

   4.   Cardiovascular System

   Acute intoxication causes hypotension, ECG changes, and cardio-vascular collapse (Stokinger 1981).

E.   Teratogenicity

   Scanlon (1975; cited by Stokinger 1981) surmised that Nb could be teratogenic because Nb crosses the placenta and because Nb catalyzes the intrinsic brain metabolite 5-hydroxytryptophan.

F.   Mutagenicity

   $NbCl_5$ was not mutagenic in the rec-assay with Bacillus subtilis (Stokinger 1981).

G.   Carcinogenicity

   There is evidence of carcinogenicity based on tests with a few Nb compounds, mainly $NbCl_5$ and the niobates (Stokinger 1981).

V.   REFERENCES

Browning E.   1969.   Toxicity of industrial metals.   2nd ed.   London: Butterworths.

Cochran KW, Doull J, Mazur M, DuBois KP.   1950.   Acute toxicity of zirconium, columbium, strontium, lanthanum, cesium, tantalum and yttrium.   Arch Ind Hyg Occup Med 1:637-650.

Merck Index, 10th ed.   1983.   Rahway, New Jersey: Merck & Co., Inc.

Smith IC, Carson BL. 1978. Trace Metals in the Environment. Volume 3-
Zirconium. Ann Arbor, MI: Ann Arbor Science Publishers, 405 pp.

Stokinger HE. 1981. Chapter 29. The metals. In: Patty's industrial
hygiene and toxicology, 3rd ed., volume IIA, Toxicology. Clayton GD, Claytor
FE, eds. New York: Wiley-Interscience, John Wiley & Sons, pp. 1493-2060.

OSMIUM

MAMMALIAN TOXICITY SUMMARY

I. INTRODUCTION

A. Occurrence and Production

Osmiridium, an alloy of platinum-group metals (24 to 80% Os), is the major natural source of osmium. The chief world source of osmiridium is South Africa. Most osmium is recovered as a by-product of copper refining in the United States. Platinum-group metals or gold placer deposit mines in Alaska, California, Oregon, and the Rocky Mountain States contribute part of U.S. osmium production. Large amounts are also toll-refined (i.e., waste or spent materials are rerefined for the owner in return for a "toll" or fee). In 1971, ∿ 2,000 troy oz (1 troy oz = 31.1 g) primary osmium and ∿ 4,200 troy oz secondary osmium were refined in the United States.

B. Uses

Few statistics exist for osmium consumption. Over a decade ago, Smith et al. (1974) surveyed users and reported that catalysis and tissue staining were the major uses for osmium tetroxide (OsO_4); osmium metal had minor uses for electrical contacts and alloys. Osmium imparts hardness to alloys used for mechanical pivots, phonograph needles, bearings, and engraving tools. Until 1969, two alloys containing 85 and 86% Os were used as fountain pen nibs. Academic and research laboratories are the major users, however. Osmic acid (=OsO_4) synovial injections are used to treat rheumatoid arthritis. Science (1981) reported a new class of potential anti-inflammatory agents based on osmium-carbohydrate polymers, which are relatively nontoxic.

C. Chemical and Physical Properties

Osmium, atomic no. 76, is the most dense metal with specific gravity of 22.48. Its compounds exhibit valences of +1 to +8. The most important compound is OsO_4, the tetroxide also called osmic acid or perosmate. OsO_4 melts at 40.6°C and boils at 131.2°C, is sparingly soluble in water, and has a pungent chlorine-like odor (Smith et al. 1978). The Os-base alloys and Os metal are so hard they must be formed by casting or by powder metallurgy (Stokinger 1981).

II. EXPOSURE AND EXPOSURE LIMITS

A. Oral

No information.

B. Inhalation

OsO_4 becomes irritating to the human eye at ~ 0.1 mg/m³ (Stokinger 1981). Environmental exposures would be expected in the United States near sites roasting and smelting copper concentrates (Smith et al. 1974). The TLV (ACGIH 1983) and permissible exposure limit (OSHA 1981) for OsO_4 is 0.002 mg Os/m³. The short-term exposure limit is 0.006 mg Os/m³.

C. Dermal

OsO_4 is a fixing and staining agent for cells and tissues. Contact dermatitis may develop.

D. Total Body Burden and Balance Information

No information.

III. TOXICOKINETICS

A. Absorption

Articular injections of osmium tetroxide are partly removed by phagocytosis (Smith et al. 1974).

B. Distribution

1. General

No information.

2. Blood

No information.

C. Excretion

No information.

IV. EFFECTS

A. Acute and Other Short-Term Exposures

Acute effects of OsO_4 in humans are a purulent discharge (Smith and Carson 1974), chiefly eye and respiratory damage. Permanent or temporary blindness has resulted. Exposed workers complain of a gritty feeling in the eyes, pain, and lacrimation and see haloes around lights. Precious-metals workers and histologists (who use 1 to 2% solution of OsO_4 as a tissue stain) suffer eye irritation, headache, asthma, or dyspnea. A fatal poisoning case from OsO_4 vapor exposure had bronchopneumonia and fatty degeneration of the renal tubule epithelium (Smith et al. 1974).   An acute immune response reaction occurred in an arthritic patient who received a second injection of osmic acid 4 yr after the first.   Some workers have experienced contact dermatitis and dermatitis develops in some arthritis patients treated with OsO_4.

A 4-h exposure to 40 ppm OsO_4 is the LC_{50} for rats and mice (Stokinger 1981).   Short-term high-level exposures to OsO_4 vapors caused purulent bronchopneumonia and cloudy swelling and granulation of the kidney tubular epithelium.   In other experiments, rabbits showed degenerative and congestive changes of other organs as well (spleen, liver, and adrenals). Exposure to the vapor or instillation of 1% OsO_4 into the conjunctival sac stained the eyes, damaged the cornea, and was followed by purulent discharge.

B.   Chronic Exposure

No information was found whether systemic effects in nonallergic patients have been ascribed to therapeutic uses of osmium.   Science (1981) implies that OsO_4 articular injections are toxic, but since they are not mobilized rapidly from the joints, the meaning probably is that handling OsO_4 before the injection could produce inhalation toxicity in medical personnel or the patient.   (Injection of osmium tetroxide solution into knee joints of patients afflicted with rheumatoid arthritis effect chemical synovectomy; that is, they destroy the synovial membrane and allow regeneration of a new thickened synovial membrane.)

C.   Biochemistry

1.   Effects on Enzymes

No information.

2.   Metabolism

Osmium injected into human joints mechanically coagulates the endothelial layer of the synovial membrane.   Some osmium remains for up to 9 mo--some is removed by phagocytosis, some remains in the cytosomes of the synovial membrane cells, and some is fixed in fat cells (Smith et al. 1974).

3.   Antagonisms and Synergisms

No information.

D.   Specific Organs and Systems

1.   Respiratory Tract

Acute high-level inhalation exposure to OsO_4 causes purulent bronchopneumonia (Smith et al. 1978).

2. Kidney

Kidney tubules are damaged after acute high-level exposure (Smith et al. 1978).

3. Skin

Dermatitis, an apparent hypersensitivity reaction, occurs in some workers and arthritis patients (Smith et al. 1978).

4. Eye

OsO_4 strongly irritates the eyes. Permanent blindness due to corneal opacity may occur. Vision is affected for several hours after exposure; the most characteristic symptom is seeing haloes around lights (Smith et al. 1978).

E. Teratogenicity

No information.

F. Mutagenicity

No information.

G. Carcinogencity

No information.

V. REFERENCES

ACGIH. 1983. TLVs[®] Threshold limit values for chemical substances and physical agents in the work environment with intended changes for 1983-84. Cincinnati, Ohio: American Conference of Governmental Industrial Hygienists.

OSHA. 1981. Occupational Safety and Health Admin. Occupational safety and health standards. Subpart 2--Toxic and hazardous substances. Code of federal regulations 29 (Part 1910.1000) pp. 673679.

Science. 1981. New ways to use metals for arthritis. Science 212:430-431.

Smith IC, Carson BL, Ferguson TL. 1974. Osmium: an appraisal of environmental exposure. Environ Health Perspect 8:201-213.

Smith IC, Carson BL, Ferguson TL. 1978. Trace metals in the environment. Vol. 4--Palladium and osmium. Ann Arbor, MI; Ann Arbor Science Publishers, 193 pp.

Stokinger HE. 1981. Chapter 29. The metals. In: Patty's industrial hygiene and toxicology. 3rd ed. Volume 2A. Clayton GD, Clayton FE, eds. New York: A Wiley-Interscience Publication. John Wiley and Sons, pp. 1493-2060.

PALLADIUM

MAMMALIAN TOXICITY SUMMARY

I. INTRODUCTION

   A. Occurrence and Production

      The platinum-group metals (Pt, Pd, Ru, Rh, Ir, and Os) are
recovered from placer deposits of two intergrown alloys of the metals and
from sulfide-ore bodies (Ni-Cu, Cu, and Cu-Co sulfides). Principal world
sources are the Bushveld Complex of South Africa, the Sudbury District of
Canada, and the Norilsk region and Kola peninsula of the USSR. Placer
deposits such as those in Alaska are minor sources. Most of the new
platinum-group metal recovery in the United States is from copper and gold
refining (Smith et al. 1978).

   B. Uses

      Since 1975, the major use of palladium in the United States has
been in automobile catalytic converters. Other catalyst uses include other
air pollution control applications, petroleum refining, and hydrogenation in
the chemical and pharmaceutical industries. Before 1975, the major use was
in telephone equipment and other electrical contacts. Palladium and its
alloys are also used in resistance windings and resistors, fuel cells,
brazing alloys, laboratory and process equipment, gas purification (mainly
hydrogen diffusion), jewelry and decorative applications, the glass industry,
dentistry, and medicine. Dental uses include toothpins in porcelain teeth,
dental wires, and gold alloys for inlays, crowns, bridges, and partial
dentures (Smith et al. 1978).

   C. Chemical and Physical Properties

      Palladium melts at 1552°C. It is soluble in HNO_3, hot concentrated
H_2SO_4, aqua regia, $HCl-HClO_3$ mixture, and fused alkalies. It is appreciably
volatile at high temperatures and is converted to PdO at red heat.

      The oxidation state of palladium in its compounds is generally
divalent. Pd(IV) and Pd(I) compounds are also known. The oxides, PdO,
$PdO_2 \cdot xH_2O$, and $Pd_2O_3 \cdot xH_2O$; the sulfide, PdS_2, and the cyanides, $Pd(CN)_4$ and
$Pd(CN)_2$, are water insoluble. Many of the simple halides and halide
complexes are water soluble. PdF_2, $PdBr_2$, and PdI_2 are insoluble in water.
$Pd(OH)_2$ dissolves in acids to give Pd(II) salts and dissolves in alkalies to
give palladites (PdO_2^{2-}).

      Palladium(II) complexes have been studied as antineoplastic drugs.
Colloidal palladium compounds have formerly been used to treat arthritis,
gout, tuberculosis, and obesity (Smith et al. 1978).

II. EXPOSURE AND EXPOSURE LIMITS

  A. Oral

      Palladium in dental alloys is innocuous (Smith et al. 1978).

  B. Inhalation

      Although attrition of Pd and Pt in the automobile catalytic converter was expected to contribute palladium to the air (Smith et al. 1978), reports of palladium in air were not found.

  C. Dermal

      No dermatitis has been reported from palladium use in jewelry.

  D. Total Body Burden and Balance Information

      No information.

III. TOXICOKINETICS

  A. Absorption

      Oral absorption is low since excretion is chiefly in the feces after ingestion (but in the urine after i.v. administration) (Smith et al. 1978). Moore et al. (1975) found < 0.5% absorption of an oral dose of ^{103}Pd (in 0.6M HCl).

  B. Distribution

      1. General

      Palladium has been found in teeth containing inlays of palladium alloys. A rabbit given a fatal i.v. dose of $PdCl_2$ had palladium in the lungs, spleen, and muscle. Rats given an i.v. injection of Na_2PdCl_4 containing ^{103}Pd had significant amounts of ^{103}Pd in the kidney, liver, and spleen at 7 days (Smith et al. 1978). After oral dosing of ^{103}PdCl$_2$, Moore et al. (1975) found Pd only in kidney and liver; but after i.v. dosing of rats, highest tissue concentrations were in kidney, spleen, liver, adrenal gland, lung, and bone. Some palladium crosses the placenta to the fetus.

      2. Blood

      Intravenous doses of ^{103}PdCl$_2$ were rapidly cleared from the blood of pregnant rats; < 0.5% of the initial dose was in the blood (compared to almost 6% in the placenta, fetus, and fetal liver) after 24 h (Moore et al. 1975). Blood of autopsied California residents did not have palladium above the detection limit (< 0.1 µg/L) (Stokinger 1981). Even palladium refinery workers had no detectable palladium in their blood (< 4 ppb) (NAS 1977).

3. Adipose

No information.

C. Excretion

Intravenous doses of $Pd(OH)_2$ or $PdCl_2$ are excreted fairly rapidly in the urine, but oral doses are largely excreted in the feces. The half-lives for palladium in kidney and liver of rats were 9 days and 6 days, respectively (Smith et al. 1978).

IV. EFFECTS

A. Acute and Other Short-Term Exposures

Serious systemic effects from palladium compounds given by subcutaneous injection, by topical application, or by ingestion have not been reported. The ineffective use of colloidal palladium to treat tuberculosis and gout was innocuous, at worst causing a feverish reaction. The usual dose in treating tuberculosis was 18 mg/day, but oral doses of 65 mg/day caused no adverse effects. Injections of colloidal $Pd(OH)_2$ (5 to 7 mg/day) to treat obesity caused necrosis at the injection site. A dilution of 1:25,000 showed a hemolytic effect. One case of contact dermatitis, a research chemist, was attributed to palladium. Acute i.v. doses of $PdCl_2$ had a slight diuretic effect in rats (Smith et al. 1978).

An i.v. dose of 1.7 mg/kg was fatal to a rabbit on the 17th day after dosing. Toxic effects included hemolysis, albuminuria, diuresis, and damage to the heart, kidneys, liver, bone marrow, and blood cells (Smith et al. 1978).

Approximate LD_{50} values of $PdCl_2$ (14 days) are 5 mg/kg (i.v., rat and rabbit), 6 mg/kg (intratracheal, rat), and 200 mg/kg (oral, rat). The acute i.v. toxicity of complex salts based on Pd are similar. Acutely poisoned animals exhibit clonic and tonic convulsions, proteinuria, and reduced water and food intake and weight. Rats given drinking water containing 92 and 194 ppm K_2PdCl_4 showed no adverse effects (Moore et al. 1975).

$(NH_4)_2PdCl_6$, $(NH_4)_2PdCl_4$, and allyl Pd chloride are severe irritants to intact skin (Stokinger 1981).

Rabbits whose skin was rubbed 7 to 8 times daily with 5.4 mg Pd HCl [sic in Chem Abstr] developed allergic dermatitis and, by the 10th day, systemic toxicity (Kolpakov et al. 1980). $PdCl_2$ was strongly allergenic to the skin of guinea pigs (Roschchin et al. 1982).

B. Chronic Exposure

Mice given 5 ppm palladium as $PdCl_2$ in drinking water for their lifetime showed reduced growth rate, but the males lived longer than the controls. Palladium was slightly carcinogenic to the mice and was definitely carcinogenic in lifetime feeding of rats (Schroeder and Mitchener 1971 and Schroeder 1970; cited by Smith et al. 1978).

Rats exposed chronically to $\geq$ 5.4 mg Pd$(NH_3)_2Cl_2$/m^3 as dust showed altered hepatic and renal functions (Panova and Veselov 1978). Bhatnagar et al. (1977) predicted, on the basis of studies in a lung organ culture system, that chronically inhaled Pd^{2+} will induce pulmonary fibrosis.

C. Biochemistry

1. Effects on Enzymes

Pd^{2+} inactivates trypsin and α-chymotrypsin, possibly by combining with free -SH or cystine groups. In fish, PdCl$_2$ at $\leq$ 2 ppm inhibited lactic dehydrogenase and glutamic oxaloacetic transaminase. Pd^{2+} binds carboxypeptidase (Smith et al. 1978). Liu et al. (1979) reported that PdCl$_2$ inhibited the following enzymes: creatinine kinase, aldolase, succinate dehydrogenase, carbonic anhydrase, alkaline phosphatase, and propyl hydroxylase. The same authors (Bhatnagar et al. 1977) also reported that PdCl$_2$ inhibited bicarbonate-stimulated ATPase. Pd^{2+} replaces the metal cofactor of many of these enzymes.

2. Metabolism

Pd(II) complexes with thiol-group containing amino acids and certain proteins (Smith et al. 1978). Pd^{2+} inhibits macromolecular synthesis. It binds to cell membranes, organelles, and proteins (Bhatnagar 1977).

3. Antagonisms and Synergisms

No information.

D. Specific Organs and Systems

Hemolysis and kidney damage appear to be commonly reported effects of acute toxicity of palladium salts and complexes.

E. Teratogenicity

PdCl$_2$ injected into chick eggs was not teratogenic (Ridgway and Karnofsky 1952; cited by Shepard 1980).

F. Mutagenicity

No information.

G. Carcinogencity

Palladium in the drinking water of mice and rats for a lifetime was carcinogenic to rats and slightly carcinogenic to mice (Smith et al. 1978).

NAS (1977) in reviewing this work by Schroeder only mentions the slight carcinogenicity in mice.

V. REFERENCES

Bhatnagar RS, Liu TZ, Lee SD. 1977. Toxicity of palladium. Presented before the Div Environ Chem, Am Chem Soc, Chicago, August 1977, 3 pp.

Kolpakov FI, Kolpakova AF, Prokhorenkov VI. 1980. Toxic and sensitizing properties of palladium hydrochloride. Gig Tr Prof Zabol 4:52-54; Chem Abstr 1980. 93(3):20277u.

Liu TZ, Lee SD, Bhatnagar RS. 1979. Toxicity of palladium. Toxicol Lett 4(6):469-473.

Moore W, Hysell D, Hall L, Campbell K, Stara J. 1975. Preliminary studies on the toxicity and metabolism of palladium and platinum. Environ Health Perspect 10:63-71.

NAS. 1977. National Academy of Sciences. Platinum group metals. Washington, DC: National Academy of Sciences Printing and Publishing Office.

Panova AI, Veselov VG. 1978. Toxicity of diamminedichloropalladium following chronic inhalation in experimental animals. Gig Tr Prof Zabol 11:45-46; Chem Abstr 1979. 90(5):34598z.

Roschin AV, Taranenko LA, Muratova NZ. 1982. Sensitizing properties of the metals indium, palladium and vanadium. Gig Tr Prof Zabol 2:5-8; Chem Abstr 1982. 96(17):137355w.

Shepard TH. 1980. Catalog of teratogenic agents, 3rd ed. Baltimore: The Johns Hopkins University Press.

Smith IC, Carson BL, Ferguson TL. 1978. Trace metals in the environment. Vol. 4 - Palladium and osmium. Ann Arbor, MI: Ann Arbor Science Publishers, 193 pp.

Stokinger HE. 1981. Chapter 29. The metals. In: Patty's industrial hygiene and toxicology. 3rd ed. Volume 2A. Clayton GD, Clayton FE, eds. New York: A Wiley-Interscience Publication. John Wiley and Sons, pp. 1493-2060.

PLATINUM

MAMMALIAN TOXICITY SUMMARY

I. INTRODUCTION

A. Occurrence and Production

The concentration of platinum in the earth's crust is 5 ppb (NAS 1977). The platinum-group metals (Pt, Pd, Ru, Rh, Ir, and Os) are recovered from placer deposits of two intergrown alloys of the metals and from sulfide-ore bodies (Ni-Cu, Cu, and Cu-Co sulfides). Principal world sources are the Bushveld Complex of South Africa, the Sudbury District of Canada, and the Norilsk region and Kola peninsula of the USSR. Placer deposits such as those in Alaska are minor sources. Most of the new platinum-group metal recovery in the United States is from copper and gold refining (Smith et al. 1978).

B. Uses

Platinum is primarily used for catalysts, e.g., in petroleum refining, automobile catalytic converters, and oxidation of NH_3 to HNO_3 or NO_x. Platinum is also used for chemically resistant laboratory and plant apparatus and vessel linings, spinnerets for extruding synthetic fibers, electrochemical anodes, and jewelry. Certain platinum complexes are used in cancer therapy (Stokinger 1981).

C. Chemical and Physical Properties

Platinum, atomic no. 78, melts at 1773.5°C and boils at 3827 ± 100°C. It dissolves in aqua regia and fused alkalies (Stokinger 1981). The chemistry of platinum is more similar to that of palladium than that of the other platinum-group metals. Representative compounds include $PtCl_4$, PtO_2, H_2PtCl_4 (chloroplatinic acid), $Na_2Pt(NO_2)_6$, $Pt(NH_3)_6Cl_4$, and $Pt(CO)Cl_2$ (Stokinger 1981).

Principal oxidation states are +2, the more common, and +4. Most platinum chemistry involves coordination compounds. H_2PtCl_6 is the product of platinum dissolution in aqua regia (NAS 1977).

II. EXPOSURE AND EXPOSURE LIMITS

A. Oral

No information.

B. Inhalation

The threshold limit value as a time-weighted average in workroom air for an 8-h day is 1 mg/m^3 for platinum metal, 0.002 mg/m^3 for soluble platinum salts (ACGIH 1983). The OSHA permissible exposure limit is 0.002 mg Pt/m³ for soluble platinum salts (OSHA 1981).

C. Dermal

Some persons exposed to platinum jewelry have exhibited allergic dermatitis (Stokinger 1981).

D. Total Body Burden and Balance Information

A total body burden is difficult to assess since platinum in tissues, if present, is usually below the detection limit (Stokinger 1981; NAS 1977).

III. TOXICOKINETICS

A. Absorption

Gastrointestinal absorption is poor for even water-soluble platinum compounds but is better after inhalation (Stokinger 1981).

B. Distribution

1. General

After inhalation of platinum metal, lung clearance is rather slow, and kidney and bone accumulate platinum. There is some transplacental passage. After intravenous administration of cis-diamminodichloroplatinum(II) platinum was highest initially in the gonads, spleen, adrenals, and excretory organs. At 6 days, a tissue-plasma ratio of 3-4:1 was maintained in the kidney, liver, ovary, and uterus. In an extensive survey of California autopsy tissues, only 4.7% contained detectable platinum at 0.003 to 1.46 µg/g (mean 0.16). After subcutaneous fat, kidney, pancreas, and liver had the highest detection frequencies (Stokinger 1981). After ingesting drinking water containing Pt^4 for 8 to 9 days, highest concentrations were in kidney and liver (Holbrook et al. 1975).

2. Blood

Although workers' urinary platinum can be detected, blood concentrations are below detection limits (NAS 1977). Rats given 319 mg Pt^4/L as $Pt(SO_4)_2 \cdot 4H_2O$ or $PtCl_4$ had mean values of 0.22 or 0.23 µg Pt/g in their blood after 8 to 9 days (Holbrook et al. 1975).

3. Adipose

As mentioned above, platinum was most frequently found in subcu-taneous fat among 16 tissues examined from California autopsies (Stokinger 1981).

C. Excretion

Excretion of absorbed platinum is about equal in urine and feces. Whole-body retention is negligible 8 days after oral dosing, since most is excreted in the feces (Stokinger 1981).

IV. EFFECTS

A. Acute and Other Short-Term Exposures

The i.v. LD_{50} in rats for $PtCl_4$ is 14.5 mg Pt/kg. The human intradermal lowest toxic dose (TD_{Lo}) is 40 mg/kg for K_2PtCl_6 and $PtCl_4$. The i.v. TD_{Lo} for cis-$Pt(NH_3)_2Cl_2$ is 2.5 mg/kg in humans.

Simple salts cause vomiting and diarrhea with bloody stools. Platinum complexes cause epileptiform convulsions, coma, and death. Non-lethal doses cause hyperirritability, restlessness, motor excitement, and delayed heart action (Stokinger 1981).

In rats, signs of acute oral toxicity also include dystrophic changes in the liver and kidneys (Veselov 1977).

B. Chronic Exposure

Platinosis, which acts like an allergic hypersensitivity that improves upon removal from exposure, is an asthma-like condition with dermatitis caused by inhaling H_2PtCl_6 or its salts at 2 to 20 μg/m³. A contact dermatitis occurs in workers exposed to oxides, chlorides, and the metal. cis-$Pt(NH_3)_2Cl_2$ causes atopic hypersensitivity, renal tubular epithelial damage (> 2 mg/kg), ototoxicity, and high-frequency hearing loss. The platinum diammine salts are immunosuppressive (Stokinger 1981). A low-grade fibrosis has been described in the lungs of some workers with platinosis (Browning 1969).

C. Biochemistry

1. Effects on Enzymes

Certain platinum complexes inhibit leucine aminopeptidase, malate dehydrogenase, liver alcohol dehydrogenase, and lactate dehydrogenase (NAS 1977).

2. Metabolism

Platinum compounds bind to DNA molecules. In the body, aquo ligands eventually replace the chloride ligands of cis-$Pt(NH_3)_2Cl_2$ and analogs (NAS 1977).

3. Antagonisms and Synergisms

A platinum complex has been reported to interfere with mito-chondrial transport of calcium (Stokinger 1981). $cis\text{-}Pt(NH_3)_2Cl_2$ inhibits DNA synthesis (NAS 1977).

D. Specific Organs and Systems

1. Respiratory System

An asthma-like syndrome called platinosis occurs after preliminary eye and upper respiratory tract irritation with cough, tightness of the chest, wheezing, and shortness of breath. The severest cases experience cyanosis, diaphoresis, feeble pulse, and clammy coldness of the extremities (Browning 1969).

2. Skin

Scaling, dryness, cracking, and eczematous patches characterize the dermatitis (Browning 1969).

E. Teratogenicity

No information.

F. Mutagenicity

$cis\text{-}Pt(NH_3)_2Cl_2$ is mutagenic in histidine-dependent Salmonella typhimurium strains TA98 and TA100 (Yasbin et al. 1980).

G. Carcinogencity

Certain platinum complexes are used as antineoplastic agents (Stokinger 1981).

V. REFERENCES

ACGIH. 1983. TLVs[®] Threshold limit values for chemical substances and physical agents in the work environment with intended changes for 1983-84. Cincinnati, Ohio: American Conference of Governmental Industrial Hygienists.

Browning E. 1969. Toxicity of Industrial Metals. 2nd ed. London: Butterworths.

Holbrook DJ Jr, Washington ME, Leake HB, Brubaker PE. 1975. Evaluation of the toxicity of various salts of lead, manganese, platinum, and palladium. Environ Health Perspect 10:95-101.

NAS. 1977. National Academy of Sciences. Platinum group metals. Washington, DC: National Academy of Sciences Printing Office.

OSHA. 1981. Occupational Safety and Health Admin. Occupational safety and health standards. Subpart 2--Toxic and hazardous substances. Code of federal regulations 29 (Part 1910.1000) pp. 673679.

Smith IC, Carson BL, Ferguson TL. 1978. Trace metals in the environment. Vol. 4 - Palladium and osmium. Ann Arbor, MI: Ann Arbor Science Publishers, 193 pp.

Snyder WS, Cook MJ, Nasset ES, Karhausen LR, Howells GP, Tipton IH. 1975. International Commission on Radiological Protection. Report of the task group on reference man. New York. ICRP Publication 23.

Stokinger HE. 1981. Chapter 29. The metals. In: Patty's industrial hygiene and toxicology. 3rd ed. Volume 2A. Clayton GD, Clayton FE, eds. New York: A Wiley-Interscience Publication. John Wiley and Sons, pp. 1493-2060.

Veselov VG. 1977. Comparative toxicity of salts containing platinum group metals following acute enteral poisoning in animals. Gig Tr Prof Zabol 7:55-57; Chem Abstr 1977. 87(15):112601z.

Yasbin RE, Matthews CR, Clarke MJ. 1980. Mutagenic and toxic effects of ruthenium. Chem-Biol Interact 31(3):355-365; Chem Abstr 1980. 93(23):215287e.

POTASSIUM

MAMMALIAN TOXICITY SUMMARY

I. INTRODUCTION

    A. Occurrence and Production

        The crustal abundance of potassium is 2.59%. Important ore min-
    erals are polyhalite, sylvite (KCl), and carnallite (KCl, $MgCl_2 \cdot 6H_2O$) (Merck
    Index 1983; Hawley 1981). In the United States, underground deposits exist
    in Carlsbad, New Mexico. Potassium salts are also recovered from brines in
    Utah and California (Minerals Yearbook 1977).

    B. Uses

        Potassium metal is used in organic syntheses. Several compounds
    have medicinal uses (Merck Index 1983). KCl and K_2SO_4 are the major salts
    produced (Minerals Yearbook 1977). KCl is used in fertilizer and plant nu-
    trients, pharmaceuticals, photography, and spectroscopy (e.g., infrared
    cells) and as a salt substitute, laboratory reagent, buffer, and food addi-
    tive (e.g., K bromate) (Merck Index 1983). The sulfate is additionally used
    in gypsum cements and in alum and glass manufacture (Hawley 1981).

    C. Chemical and Physical Properties

        Potassium, atomic no. 19, melts at 63.2°C and boils at 765.5°C. It
    reacts vigorously with oxygen and with water even at -100°C. It exhibits a
    valence of +1 in its compounds (Merck Index 1983).

II. EXPOSURE AND EXPOSURE LIMITS

    A. Oral

        Natural potassium content and food additives contribute ∿ 3.3 g K/
    day to the daily intake (Snyder et al. 1975). NAS (1980) estimated that
    daily intake by adults is 1,950 to 5,900 mg. Meat, milk, and many fruits are
    good potassium sources.

    B. Inhalation

        No information.

    C. Dermal

        No information.

D. Total Body Burden and Balance Information

The potassium balance for 70-kg reference man, according to Snyder et al. (1975), is:

Intake, mg/day	Losses, mg/day	
Food and fluids 3.3	Urine	2.8
	Feces	0.36
	Sweat	0.13
	Other fluids	trace

The total body burden is 140 g, 120 g of which is in soft tissue.

III. TOXICOKINETICS

A. Absorption

Gastrointestinal absorption is > 90% (NAS 1980).

B. Distribution

1. General

Of the 140-g body burden, 84 g is in muscle and 15 g is in skeleton. Potassium also accumulates in fat, blood, central nervous system, intestines, liver, lung, and skin in gram quantities (Snyder et al. 1975). K^+ is 30 times more concentrated in intracellular than extracellular fluids (Harrow and Mazur 1962).

2. Blood

Wide variations in intake do not affect blood levels because of homeostatic control mechanisms (NAS 1980). Of the 8.8 g total K in blood, 8.3 g is in the red cells and 0.5 g is in the plasma (Snyder et al. 1975). Blood concentrations of potassium are increased in adrenal cortex deficiency (Addison's disease) (Harrow and Mazur 1962).

Normal adult values for serum are 3.6 to 5.5 mEq/L (140 to 215 mg/L). Going from a cold to a warm climate increases blood potassium by $\leq$ 30% (Henry 1964).

Numerous sampling and sample handling precautions must be followed to assure accurate K^+ in blood determinations. Opening and closing the fist while a tourniquet is applied may increase serum K+ 10 to 20% for 2 min. Of course, anticoagulants containing K^+ cannot be used. Hemolysis must be avoided since erythrocytes contain ~ 20 times as much K^+ as the serum or plasma, and erythrocytes should be removed to prevent changes in serum or plasma concentrations due to active transport into the erythrocytes (Henry 1964).

3. Adipose

Snyder et al. (1975) give 4.8 g as the total K content of adipose tissue with 2.4 g in subcutaneous fat, 1.6 g in other separable fat, and 0.26 g in interstitial fat.

C. Excretion

Potassium in the body is under homeostatic control and regulated by the kidney (NAS 1980).

IV. EFFECTS

A. Acute and Other Short-Term Exposures

Sudden increases above $\sim$ 18 g/day for an adult (12.0 g/m^2 of surface area) cause acute poisoning, which may produce cardiac arrest (NAS 1980).

Abnormally high potassium concentrations in the blood arise most often from defective excretion by the kidney. The conditions is characterized by electrocardiograph abnormalities; weakness and flaccid paralysis arise in severe cases. The ECG changes include elevated T waves, depressed P waves, and eventually atrial asystole (Dorland 1974). One of every 400 patients receiving 40 to 90 mEq K^+/day in a Boston hospital died of sudden cardiac arrest. These patients had unrecognized diminished kidney function (Kewitz 1974).

B. Chronic Exposure

Variations of the Na:K ratio in the diet may affect blood pressure (NAS 1980).

C. Biochemistry

1. Effects on Enzymes

K^+ activates glycerol dehydrogenase, mitochondrial pyruvate carboxylase, pyruvate kinase, vitamin B_{12}-dependent diol dehydratase, L-threonine dehydratase, adenosine triphosphatase, aminoacyl transferase, and probably numerous other enzymes (Smith and Carson 1977).*

---

* None of the secondary sources consulted gave much information on the physiological function of potassium. Because Tl^+ mimics the behavior of K^+ in the body (in fact, Tl^+ binds to many ligands $\geq$ 10 times stronger than K^+ and activates enzymes to a greater extent than K^+), the thallium review by Smith and Carson (1977) was useful.

2. Metabolism

K^+ in enzyme probably binds at monoanion phosphate and carboxylate centers whose "hole" size closely matches the ionic radius of K^+. Proton nuclear magnetic resonance studies show that K^+ is bound in polyphosphate chelates in pyrophosphate, adenosine, phosphate, and adenosine triphosphate complexes (Smith and Carson 1977).

3. Antagonisms and Synergisms

Increasing the ingestion of K^+ increases the urinary Na^+ excretion and vice versa (Anderson 1953).

4. Physiological Requirements

NAS (1980) estimated that 1,875 to 5,625 mg K/day is a safe and adequate daily dietary intake for potassium.

D. Specific Organs and Systems

No information.

E. Teratogenicity

No teratogenic effects have been reported due to excess potassium. Potassium deficiency does not account for limb defects produced by administration of the carbonic anhydrase inhibitor acetazolimide (Wilson et al. 1968; cited by Shepard et al. 1980), but deficient K^+ in an organ culture of fetal mouse kidney caused abnormal branching of tubules and cystic dilatations of the ureteral buds (Crocker and Vernier 1970; cited by Shepard et al. 1980).

F. Mutagenicity

No information.

G. Carcinogenicity

No information.

V. REFERENCES

Anderson AK. 1953. Essentials of physiological chemistry. 4th ed. New York, NY: John Wiley and Sons, Inc.

Dorland. 1974. Dorland's illustrated medical dictionary, 25th ed. Philadelphia, PA: W. B. Saunders.

Henry RJ. 1964. Clinical chemistry: principles and technics. New York, NY: Hoeber Medical Division, Harper & Row Publishers.

Kewitz H.   1974.   Arzneimittel als Krankheitsursache [Drugs as causes of illness].   Internist (Springer-Verlag) 15(1):7-12.

Minerals Yearbook 1975.   1977.   Volume I.   Metals, minerals, and fuels. Bureau of Mines.   U.S. Department of the Interior.   Washington DC: U.S. Government Printing Office.

Shepard TH.   1980.   Catalog of teratogenic agents.   3rd ed.   Baltimore: The Johns Hopkins University Press.

Smith IC, Carson BL.   1977.   Trace metals in the environment: volume 1 - thallium.   Ann Arbor, Michigan: Ann Arbor Science Publishers, Inc., pp. 394.

NAS.   1980.   National Academy of Sciences.   Recommended dietary allowances. 9th ed.   Washington, DC: Printing and Publishing Office, National Academy of Sciences.

Snyder WS, Cook MJ, Nasset ES, Karhausen LR, Howells GP, Tipton IH.   1975. International Commission on Radiological Protection.   Report of the task group on reference man.   New York.   ICRP Publication 23.

Stokinger HE.   1981.   Chapter 29.   The metals.   In Patty's industrial hygiene and toxicology, 3rd ed., volume IIA, Toxicology.   Clayton GD, Clayton FE, eds.   New York: Wiley-Interscience, John Wiley and Sons.   pp. 1493-2060.

RHENIUM

MAMMALIAN TOXICITY SUMMARY

I. INTRODUCTION

A. Occurrence and Production

The crustal abundance of rhenium is 0.001 ppm. It occurs in nature in gadolinite, molybdenite, columbite, rare earth minerals, and some sulfide ores (Merck Index 1983). Rhenium has been recovered from molybdenum and copper concentrates. It is extracted from molybdenum concentrates as K perrhenate, which is reduced to the metal (Browning 1969).

B. Uses

The industrial uses of rhenium include contacts in marine engine magnets, electron tubes, and heater elements for metal evaporation (Browning 1969). Merck Index (1983) also lists plating and possible use in high-temperature thermocouples and in tungsten and molybdenum alloys.

C. Chemical and Physical Properties

Rhenium, atomic no. 75, melts at 3180°C and boils at 5900°C. Its specific gravity is 21.2. The metal is resistant to salt water and hydro-chloric acid. Although sulfuric acid has little effect, it is dissolved by nitric acid and other oxidizing mineral acids. Its chemistry is similar to that of ^{99}Tc (Browning 1969). Rhenium exhibits valences 1 to 7, but the +7 state is most stable. Of the two natural isotopes ^{185}Re and ^{187}Re, the latter accounts for 62.93% and is radioactive (half-life of $\sim 10^{11}$ y). Re_2O_7 is freely soluble in water to give perrhenic acid, $HReO_4$. ReO_3 is practically insoluble in water, alkalies, and nonoxidizing acids. Nitric acid oxidizes ReO_3 to $HReO_4$ (Merck Index 1983). Rhenium trihalides dissolve in water, the tetrahalides decompose on dissolution, and the pentahalides hydrolyze. Complex halides are also formed, e.g., K_2ReCl_6 (Luckey et al. 1975).

II. EXPOSURE AND EXPOSURE LIMITS

No information.

III. TOXICOKINETICS

A. Absorption

Luckey et al. (1975) conjectured that Re absorption should be the same as that as manganese or greater.

B. Distribution

1. General

^{188}Re accumulates in the thyroid more than in other tissues, presumably by selective filtration from the blood. When rhenium was injected
s.c. followed in 1 h by a tracer dose of ^{188}Re, the only significant amounts
were found in the thyroid and gastrointestinal tract (Browning 1969). Luckey
et al. (1975) describe the same tests, stating that the rats tested were deficient in iodine or had been made goiterous by thiouracil. An intravenous
injection of NaReO$_4$ was retained transiently in small quantities in bone and
liver (Luckey et al. 1975).

2. Blood

No information.

C. Excretion

After an injection of ^{188}Re nitrate, > 90% was recovered in the
urine within 24 h (Browning 1969). An intravenous injection of ^{183,184}Rd as
NaReO$_4$ was also excreted mainly in the urine. About 13% of an i.v.
administered dose of ^{183}Re was found in the digestive tract (Luckey
et al. 1975).

IV. EFFECTS

A. Acute and Other Short-Term Exposures

Maresh et al. (1940; cited by Haley and Cartwright 1968) reported
on i.p. lethal range of 900 to 1,000 mg Re/kg for NaReO$_4$ in rats, which
developed cyanosis, increased respiration, and tonic convulsions. Re$_2$Cl$_6$ was
more toxic but an LD$_{50}$ was not determined. Both caused transient tachycardia
and slight hypotension in dogs.

The i.p. LD$_{50}$ in 7 days for male CF1 mice was 2,800 mg/kg for KReO$_4$
and 280 mg/kg for Re$_2$Cl$_6$. Mice dosed with KReO$_4$ showed sedation and severe
ataxia; those dosed with Re$_2$Cl$_6$, sedation and abdominal irritation. Intravenous doses of 10 to 50 mg/kg of KReO$_4$ produced tachycardia and transient
hypertension in cats; at 60-70 mg/kg, hypotension, bradycardia, and bradypnea.
At 70 mg/kg, 4 of 5 cats died from cardiovascular collapse and respiratory
failure. Similar studies in cats with Re$_2$Cl$_6$ produced slight hypertension at
10 mg/kg and death at 20 mg/kg, effects that were entirely due to the release
of hydrochloric acid.

See Part IV.D.1 for skin and eye irritation.

B. Chronic Exposure

No information.

C. Biochemistry

1. Effects on Enzymes

No information.

2. Metabolism

When given i.p. to rats, K_2ReCl_6 and $ReCl_3$ are hydrolyzed to Re oxides (Luckey et al. 1975). The latter releases HCl (Haley and Cartwright 1968).

3. Antagonisms and Synergisms

In the cat experiments of Haley and Cartwright (1968), $KReO_4$ did not block the pharmacologic responses to acetylcholine, epinephrine, or histamine and atropinization did not counteract the effects of $KReO_4$ on the heart.

D. Specific Organs and Systems

1. Skin

$KReO_4$ and Re_2Cl_6 at 0.5 g did not irritate abraded or unabraded rabbit skin. Re_2Cl_6 caused a permanent black stain on rabbit and human skin (Haley and Cartwright 1968).

2. Eye

$KReO_4$ did not damage rabbit eyes in the Draize test. Re_2Cl_6 produced effects reversible within 24 h: immediate corneal irritation, congestion and swelling of the iris, redness of the conjuctiva, and slight lacrimation (Haley and Cartwright 1968).

E. Teratogenicity

No information.

F. Mutagenicity

No information.

G. Carcinogenicity

No information.

V. REFERENCES

Browning E. 1969. Toxicity of industrial metals. 2nd ed. London: Butterworths.

Haley TJ, Cartwright FD.   1968.   Pharmacology and toxicology of potassium perrhenate and rhenium trichloride.   J Pharm Sci 57:321-323.

Luckey TD, Venugopal B, Hutcheson D.   1975.   Environ Qual Saf Suppl., 1 (Heavy Metal Toxicity Safety and Hormology):1-120.

Merck Index, 10th ed.   1983.   Rahway, New Jersey: Merck & Co., Inc.

RHODIUM

MAMMALIAN TOXICITY SUMMARY

I. INTRODUCTION

 A. Occurrence and Production

 The platinum-group metals (Pt, Pd, Ru, Rh, Ir, and Os) are
recovered from placer deposits of two intergrown alloys of the metals and
from sulfide-ore bodies (Ni-Cu, Cu, and Cu-Co sulfides). Principal world
sources are the Bushveld Complex of South Africa, the Sudbury District of
Canada, and the Norilsk region and Kola peninsula of the USSR. Placer
deposits such as those in Alaska are minor sources. Most of the new
platinum-group metal recovery in the United States is from copper and gold
refining (Smith et al. 1978).

 B. Uses

 Rhodium is used as a catalyst (alloyed with platinum to oxidize NH_3
to HNO_3 or NO_x), in electrical contacts, electroplating of jewelry and
instruments, and in glass (e.g., reflectors in searchlights and movie
projectors. Rh-Pt alloys are used for crucibles, spinnerets, resistance
furnace windings, and thermocouples (Stokinger 1981; Browning 1969).

 C. Chemical and Physical Properties

 Rhodium, atomic no. 45, melts at 1965°C and boils at 2500°C. Its
electrodeposited coatings are very hard and more corrosion resistant than
platinum (Browning 1969). The chemistry of rhodium resembles that of
iridium. Both form protective oxide coatings that decompose at high
temperatures, but the rhodium oxide dissociation is much slower. Rhodium
compounds exhibit only the +3 oxidation state. Principal compounds include
Rh_2O_3, $RhCl_3$, H_3RhCl_6, $Na_3Rh(NO_2)_2$, $[Rh(NH_3)_6]_2Cl_6$, $Rh(CO)_4$, $Rh(CO)_3$, and
$Rh(CO)_2Cl$ (Stokinger 1981).

II. EXPOSURE AND EXPOSURE LIMITS

 A. Oral

 No information.

 B. Inhalation

 The threshold limit value (TLV) as a time-weighted average in
workroom air for an 8-h day is 0.001 mg/m³ for soluble rhodium salts; the
short-term exposure limit is 0.003 mg/m³/15 min. The TLV for rhodium metal
fume and dusts is 1 mg/m³ (ACGIH 1983). The OSHA permissible exposure limit
is 0.001 mg/m³ for soluble rhodium salts, 0.1 mg Rh/m³ for rhodium metal fume
and dusts (OSHA 1981).

C. Dermal

Rhodium is apparently not an allergen (Browning 1969).

D. Total Body Burden and Balance Information

No information.

III. TOXICOKINETICS

A. Absorption

Most of the rhodium metal and rhodium dioxide (particle size 0.07 to 0.12 μm) inhaled by dogs remained in the lungs > 3 yr after exposure. The rest was mainly in lymph nodes (Stokinger 1981).

B. Distribution

1. General

No information.

2. Blood

No information.

3. Adipose

No information.

C. Excretion

No information.

IV. EFFECTS

A. Acute and Other Short-Term Exposures

The i.v. LD_{50} of $RhCl_3$ for rats is 198 mg/kg; for rabbits, 215 mg/kg. Rats died within 48 h; survivors did not show any histological lesions in selected organs after 100 days. For rabbits these times were 12 h and 30 days, respectively (Stokinger 1981).

B. Chronic Exposure

No information.

C. Biochemistry

  1. Effects on Enzymes

No information.

  2. Metabolism

No information.

  3. Antagonisms and Synergisms

The effectiveness of $RhCl_3$ as an antiviral chemotherapeutic has been variously explained by an ability to act as a cobalt antagonist and by the ability to form lipoid-soluble complexes, which interferes with phospholipid formation by the virus (Browning 1969).

D. Specific Organs and Systems

No information.

E. Teratogenicity

Chick embryos exposed on the 8th day of incubation to Rh chloride were stunted; mild micromelia (reduction of limb size) and feather growth inhibition were also observed (Ridgway and Karnofsky 1952; cited by Shepard 1980).

F. Mutagenicity

No information.

G. Carcinogencity

No information.

V. REFERENCES

ACGIH.   1983.   TLVs[®] Threshold limit values for chemical substances and physical agents in the work environment with intended changes for 1983-84. Cincinnati, Ohio: American Conference of Governmental Industrial Hygienists.

Browning E.   1969.   Toxicity of Industrial Metals.   2nd ed.   London: Butterworths.

OSHA.   1981. Occupational Safety and Health Admin.   Occupational safety and health standards.   Subpart 2--Toxic and hazardous substances.   Code of federal regulations 29 (Part 1910.1000) pp. 673679.

Shepard TH. 1980. Catalog of teratogenic agents, 3rd ed. Baltimore: The Johns Hopkins University Press.

Smith IC, Carson BL, Ferguson TL. 1978. Trace metals in the environment. Vol. 4 - Palladium and osmium. Ann Arbor, MI: Ann Arbor Science Publishers 193 pp.

Stokinger HE. 1981. Chapter 29. The metals. In: Patty's industrial hygiene and toxicology. 3rd ed. Volume 2A. Clayton GD, Clayton FE, eds. New York: A Wiley-Interscience Publication. John Wiley and Sons, pp. 1493-2060.

RUBIDIUM

MAMMALIAN TOXICITY SUMMARY

I. INTRODUCTION

A. Occurrence and Production

Rubidium occurs in the earth's crust in amounts of 310 ppm and in seawater at 0.2 ppm. It is found in the minerals carnallite and beryl and in iron ores. Rubidium is producted as a by-product of potassium production from carnallite (Browning 1969).

B. Uses

It is used in the manufacture of photoelectric cells and in making zeolite catalysts (Merck Index 1983).

C. Chemistry

Rubidium, atomic no. 37, melts at 39°C, boils at 700°C. Rubidium has chemical and physical properties expected of an alkali metal. It oxidizes rapidly in air and decomposes water with liberation of hydrogen (Browning 1969). RbOH is a stronger base than KOH (Merck Index 1983).

II. EXPOSURE AND EXPOSURE LIMITS

A. Oral

Estimates of 2.5 mg Rb/day for Japanese adult males and 1.8 mg Rb/day for adult females were made for rubidium intake from food (Snyder et al. 1975). A value of 2.5 mg/day was also reported by Clemente et al. (1977) for the median daily intake by the Italian population.

B. Inhalation

No information was found in secondary sources.

C. Dermal

No information.

D. Total Body Burden and Balance Information

Normal total body burden of rubidium for a 70-kg man has been estimated to be 360 mg (Lang 1935; cited by Stokinger 1981). Snyder et al. (1975) estimated 680 mg (470 mg in soft tissue). According to Snyder et al. (1975), the rubidium balance for 70-kg reference man [values by Clemente et al. (1977) are in brackets] is:

Intake, mg/day	Losses, mg/day
Foods and fluids 2.2 [2.5]	Urine 1.9 [2.0]
	Feces 0.3 [1.5]
	Sweat and other fluids 0.05

III. TOXICOKINETICS

A. Absorption

Presumably all ingested rubidium is absorbed (Snyder et al. 1975). Most animal tissues contain 20 to 40 ppm. Rubidium is treated very similarly to potassium by the body and will replace a certain percentage of potassium when available in the body (Stokinger 1981).

B. Distribution

1. General

Rubidium is stored in muscles and major organs such as liver, spleen, lung, kidney, heart, and brain (Stokinger 1981).

2. Blood

Rubidium is quickly distributed by the blood to other tissues with some being retained by the red cells (Stokinger 1981). Snyder et al. (1975) estimated 14 mg Rb in whole blood, 2 mg in plasma and 12 mg in the red cells.

C. Excretion

According to some sources, rubidium is excreted mainly by the kidneys. Others report data indicating primarily fecal excretion (Snyder et al. 1975). It has been estimated that the biological half-life of rubidium in man is 50 to 60 days based on urinary excretion rates (Fieve 1971; cited by Stokinger 1981).

IV. EFFECTS

A. Acute and Other Short-Term Exposures

The toxicity of rubidium appears to be relatively low; the oral LD_{50} of rubidium chloride in mice was 3800 mg/kg. The intraperitoneal LD_{50} was 1160 mg/kg, and for rats, the intraperitoneal LD_{50} was 1200 mg/kg. The oral LD_{50} for the hydroxide was 586 mg/kg for the mouse and 900 mg/kg for the rat. Toxic doses of the strongly basic hydroxide resulted in gastrointestinal hemorrhages with thinning and necrotization of the walls, adhesions of abdominal organs, and initial hyperexcitability followed by weakness. Rubidium hydroxide caused mild skin irritation (Stokinger 1981).

B.   Chronic Exposure

In a three-generation study, rats were given 121 mg RbCl/L and 75 mg KCl/L in their drinking water (the diet also contained 0.89% K) The only adverse effect noted was greater excitability, mild cannibalism, and a greater startle response compared to the controls.   To determine if the rubidium/potassium intake ratio was important, the rats were put on a potassium-deficient diet containing the same rubidium levels and the rats experienced seizures and had 42% mortality.   The survivors put on a normal diet returned to health.   In this study, the rubidium intake was 10% of the potassium intake and this may be the upper limit compatible with health (Meltzer and Lieberman 1971; cited by Stokinger 1981).

C.   Biochemistry

1.   Effects on Enzymes

Rb^+ appears to be transported into brain, red cells, and muscles by the same active transport mechanism as for K^+ (Na^+,K^+-adenosine triphosphatase) (Stokinger 1981).

2.   Metabolism

Rubidium has a close similarity to potassium in the body and can replace potassium in certain physiological processes.   It functions similarly to potassium in altering heart muscle contractions and can alter behavior and manic-depressive states, but its metabolic function is not understood (Stokinger 1981).

3.   Antagonisms and Synergisms

Rubidium salts interfere with thyroid uptake of iodine (Stokinger 1981).

4.   Physiological Requirements

It is present in the body in larger than trace metal amounts and it can replace potassium in certain processes, but the body's actual requirement for rubidium is not known (Stokinger 1981).

D.   Specific Organs and Systems

Rubidium is known to affect the brain in altering behavior and manic-depressive states; it appears to act by altering basic metabolic functions (Stokinge 1981).

E.   Teratogenicity

No information.

F.   Mutagenicity

No information.

G.   Carcinogenicity

No information.

V.   REFERENCES

Browning E. 1969. Toxicity of industrial metals. 2nd ed. London:
Butterworths.

Clemente GF, Rossi LC, Santaroni GP. 1978. Studies on the distribution and
related health effects of the trace elements in Italy. In: Trace substances
in environmental health XII. Hemphill DD, Ed. Proceedings of University of
Missouri's 12th Annual Conference on Trace Substances in Environmental
Health, Columbia, Missouri, June 6-8, 1978, University of Missouri, Columbia,
Missouri, pp. 23-30.

Merck Index, 10th ed. 1983. Rahway, New Jersey: Merck & Co., Inc.

Snyder WS, Cook MJ, Nasset ES, Karhausen LR, Howells GP, Tipton IH. 1975.
International Commission on Radiological Protection. Report of the task
group on reference man. New York. ICRP Publication 23.

Stokinger HE. 1981. Chapter 29, The metals. In: Patty's industrial
hygiene and toxicology, 3rd ed., volume IIA, Toxicology. Clayton GD, Clayton
FE, eds. New York: Wiley-Interscience, John Wiley & Sons, pp. 1493-2060.

RUTHENIUM

MAMMALIAN TOXICITY SUMMARY

I. INTRODUCTION

A. Occurrence and Production

The platinum-group metals (Pt, Pd, Ru, Rh, Ir, and Os) are re-covered from placer deposits of two intergrown alloys of the metals and from sulfide-ore bodies (Ni-Cu, Cu, and Cu-Co sulfides). Principal world sources are the Bushveld Complex of South Africa, the Sudbury District of Canada, and the Norilsk region and Kola peninsula of the USSR. Placer deposits such as those in Alaska are minor sources. Most of the new platinum-group metal recovery in the United States is from copper and gold refining (Smith et al. 1978).

B. Uses

Ruthenium is used as a hardener for platinum and palladium alloys for jewelry and electrical contact (Browning 1969).

C. Chemical and Physical Properties

Ruthenium, atomic no. 44, melts at 2310°C and boils at 3900°C. It dissolves in fused alkalies and is slightly soluble in aqua regia. The chemistry of ruthenium is more similar to that of osmium than of the other platinum-group metals. Both form dissociable volatile oxides at high tem-peratures. Compounds exhibit positive oxidation states of 1, 3, 4, 5, 6, 7, and 8 (Stokinger 1981).

II. EXPOSURE AND EXPOSURE LIMITS

A. Oral

No information.

B. Inhalation

Akinfieva (1981) recommended a maximum permissible concentration of 1 mg RuO_2/m^3 in workroom air. For $RuOHCl_2$, the value should be 0.1 mg/m^3 (Akinfieva 1979).

C. Dermal

No information.

D. Total Body Burden and Balance Information

No information.

III. TOXICOKINETICS

    A. Absorption

        Absorption of ruthenium compounds is < 8% by ingestion (Stokinger 1981).

    B. Distribution

        1. General

        After an i.p. dose of ^{106}RuCl$_3$, dogs showed more ^{106}Ru in muscle and bone at 70 days than in liver and kidneys (Stokinger 1981).

        2. Blood

        No information.

        3. Adipose

        No information.

    C. Excretion

        Most ruthenium is excreted in the feces (Stokinger 1981).

IV. EFFECTS

    A. Acute and Other Short-Term Exposures

        A Ru^{3+} ammine complex had an LD$_{50}$ (route not given) of 132 mg/kg. The oxide, like that of osmium, is highly injurious to eyes and lungs (Browning 1969; Stokinger 1981).

        The i.p. LD$_{50}$'s were determined by Kruszyna et al. (1980) for a series of ruthenium compounds: Ru(NO)(NH$_3$)$_5$Cl$_2$, 8.9; chloronitrosylbis(2,2'-dipyridyl)ruthenium(II), 55; dichlorobis(2,2'-dipyridyl)ruthenium(II), 63; RuCl$_3$, 108; and K$_3$[Ru(NO)Cl$_5$], 124 mg/kg. Toxic effects appeared to be due to ruthenium and included convulsions and respiratory arrest.

        The LD$_{50}$ of RuO$_2$ in mice was 5,570 mg/kg, respectively (presumably oral dosing). In rats, the LD$_{50}$'s for RuO$_2$ and RuOHCl$_2$ were 4,580 and 1,250 mg/kg, respectively (Atkinfieva and Bubnova 1978). Oral LD$_{50}$ doses of Ru(OH)Cl$_2$ in mice and guinea pigs were 462.5 and 402.5 mg/kg, respectively. By inhalation, 21.4 mg RuO$_2$/m^3 was quite toxic, causing morphological and functional changes, especially in the liver (Akinfieva 1981).

    B. Chronic Exposure

        Ruthenium red, Ru$_2$(OH)$_2$Cl$_4$·7NH$_3$·3H$_2$O, is a tumor inhibitor apparently by interfering with mitochondrial transport of calcium (Stokinger 1981)

Chronic [oral?] dosing of rats with RuO_2 caused changes in the cardiovascular system, lipid metabolism, and acid-base equilibrium (Akinfieva and Bubnova 1978).

C. Biochemistry

1. Effects on Enzymes

Acute inhalation of RuO_2 by rats caused an increase in the activity of alkaline phosphatase (Akinfieva 1981).

2. Metabolism

No information.

3. Antagonisms and Synergisms

No information.

D. Specific Organs and Systems

RuO_2 is a strong respiratory irritant.

E. Teratogenicity

No information.

F. Mutagenicity

Ruthenium complexes are mutagenic, causing Salmonella typhimurium strains TA98 and TA100 to revert to histidine independence (Yasbin et al. 1980)

G. Carcinogencity

No information.

V. REFERENCES

Akinfieva TA, Bubnova NI. 1978. Experimental data on the effect of ruthenium compounds on the body. In: Aktual Probl Gig Tr, Tarasenko N Yu, Ed. Moscow, USSR: Pervyi Moskovskii Med. Inst.; Chem Abstr 1979 91(15):118213d.

Akinfieva TA. 1979. Action of ruthenium hydroxychloride on an organism. Gig Tr Prof Zabol 6:54-55.

Akinfieva TA. 1981. Basis for the maximum allowable concentration of ruthenium dioxide in the air of work areas. Gig Tr Prof Zabol 1:46-47; Chem Abstr 1981 95(3):19077p.

Browning E. 1969. Toxicity of Industrial Metals. 2nd ed. London: Butterworths.

Kruszyna H, Kruszyna R, Hurst J, Smith RP. 1980. Toxicology and pharmacology of some ruthenium compounds: vascular smooth muscle relaxation by nitrosyl derivatives of ruthenium and iridium. J Toxicol Environ Health 6(4):757-773; Chem Abstr 1980. 93(25):230737u.

Smith IC, Carson BL, Ferguson TL. 1978. Trace metals in the environment. Vol. 4 - Palladium and osmium. Ann Arbor, MI: Ann Arbor Science Publishers, 193 pp.

Stokinger HE. 1981. Chapter 29. The metals. In: Patty's industrial hygiene and toxicology. 3rd ed. Volume 2A. Clayton GD, Clayton FE, eds. New York: A Wiley-Interscience Publication. John Wiley and Sons, pp. 1493-2060.

Yasbin RE, Matthews CR, Clarke MJ. 1980. Mutagenic and toxic effects of ruthenium. Chem-Biol Interact 31(3):355-365; Chem Abstr 1980. 93(23): 215187e

SCANDIUM

MAMMALIAN TOXICITY SUMMARY

I. INTRODUCTION

    A. Occurrence and Production

        Scandium occurs in the earth's crust at 5 to 6 ppm. This metal,
    found with and sometimes classified with the rare earth metals, is widely
    dispersed in nature. It occurs in the mineral thortveitite [$(Sc,Y)_2Si_2O_7$]
    and other rare earth minerals such as davidite, ytterbite, orthite, and
    cerrite. It is also frequently found associated with tin or zirconium (Merck
    Index 1983).

    B. Uses

        Research and development are the major scandium uses.
    Industrially, the metal has been used in special high-intensity mercury vapor
    lamps for outdoor color television lighting and in magnesium alloys (Minerals
    Yearbook 1977).

    C. Chemistry

        Scandium, atomic no. 21, melts at 1538°C. Salts are hydrolyzed in
    aqueous solution to $Sc(OH)_3$. Other common compounds include Sc_2O_3 (scandia),
    $ScCl_3$, $Sc_2(SO_4)_3$, and $Sc(NO_3)_3$ (Merck Index 1983).

II. EXPOSURE AND EXPOSURE LIMITS

    A. Oral

        No data were found for United States food and drinking water.
    Clemente (1976) tabulated scandium concentrations in Italian drinking waters:
    < 0.003 to 0.1 µg/L (mean < 0.01 µg/L). Clemente (1976) gave < 0.02 µg as
    the average daily intake in Italy from drinking water and 0.2 µg from the
    cooked diet.

    B. Inhalation

        Scandium in the air apparently has no anthropogenic sources.
    Scandium is one of the elements used to determine enrichment factors for
    anthropogenic elements in air as compared to the crustal concentration.
    Analyses are by neutron activation (Smith and Carson 1980). The minimim,
    detectable amount of scandium in air by NAA is 0.005 ng/m^3. The estimated
    average daily intake in Italy through inhalation is 0.04 µg (Clemente 1976).

    C. Dermal

        No information.

D.   Total Body Burden and Balance Information

Clemente (1976) gave the following balance information for the Italian population: dietary intake 0.2 µg, which was practically all excreted in the feces; < 0.01 µg was found in the daily urine. Clemente et al. (1977) gave the mean dietary intake as 0.17 µg, but the median fecal value was higher: 0.4 µg/day.

III.   TOXICOKINETICS

A.   Absorption

Scandium chloride is very poorly absorbed from the gastrointestinal tract. It is used as a marker to measure intestinal absorption since fecal recovery is 98% (Luckey et al. 1975).

B.   Distribution

1.   General

From intravenously administered scandium, most was found in the liver and reticuloendothelial system with small amounts in the bone (Luckey et al. 1975).

2.   Blood

Clemente (1976) and Clemente et al. (1977) published a scandium concentration in blood of the Italian population as < 0.03 ng/mL (< 0.03 µg/L) determined by neutron activation analysis.

C.   Excretion

The greatest excretion from all forms of intake is fecal with traces excreted in the urine (Luckey et al. 1975).

IV.   EFFECTS

A.   Acute and Other Short-Exposures

The oral LD_{50} of scandium chloride in the mouse was 690 mg/kg, the intraperitoneal LD_{50} was 130 mg/kg. Oral toxicity contradicts poor absorption; the toxicity is perhaps due to mass effect, maybe the rupture or blocking of the gastrointestinal tract due to the accumulation of the poorly absorbed or poorly metabolized scandium chloride (Luckey et al. 1975).

Another study found low toxicity for scandium chloride; the oral LD_{50} in male mice was 755 mg/kg; the intraperitoneal LD_{50} was 4000 mg/kg (Haley 1962; cited by Haley 1965).

It has been reported that inhalation of rare earth oxide mixtures including scandium by guinea pigs resulted in fatal delayed chemical hyperemia and cellular eosinophilia. An isolated cellular vascular granuloma occurred after 1 yr in surviving animals (Haley 1965).

Scandium has an irritating effect to the eyes producing conjunctival ulcers. It also causes extensive damage to injured or abraded skin resulting in epilation and scar formation but causes no irritation to intact skin. Skin lesions develop from intradermal injection in guinea pigs and humans (Haley 1965).

B. Chronic Exposure

Scandium caused no internal organ damage when fed at levels of 0.01, 0.1, and 1% of the diet of rabbits for 90 days. There was also no effect on growth (Haley 1965).

C. Biochemistry

1. Effects on Enzymes

Scandium and the other rare earths act as anticoagulants (Haley 1965).

2. Metabolism

No information.

3. Antagonisms and Synergisms

The rare earths including scandium are antagonists of thrombokinase (Haley 1965).

D. Specific Organs and Systems

Scandium when administered intravenously to animals decreased blood pressure; in cats and dogs, hypotension was produced followed by death from cardiovascular collapse coupled with respiratory paralysis (Haley 1965).

E. Teratogenicity

No information.

F. Mutagenicity

No information.

G. Carcinogenicity

No carcinogenic effects reported, but skin and lung granulomas have been induced after local ingestion or inhalation (Haley 1965).

V. <u>REFERENCES</u>

Clemente GF. 1976. Trace element pathways from environment to man.
J Radioanal Chem 32(1):25-41.

Clemente GF, Rossi LC, Santaroni GP. 1978. Studies on the distribution and
related health effects of the trace elements in Italy. In: Trace substances
in environmental health XII. Hemphill DD, Ed. Proceedings of University of
Missouri's 12th Annual Conference on Trace Substances in Environmental
Health, Columbia, Missouri, June 6-8, 1978, University of Missouri, Columbia,
Missouri, pp. 23-30.

Haley TJ. 1965. Pharmacology and toxicology of the rare earth elements.   J
Pharm Sci 54:663-670.

Luckey TD, Venugopal B, Hutcheson D. 1975. Environ Qual Saf Suppl, 1 (Heavy
Metal Toxicity Safety and Hormology):1-120.

Merck Index, 10th ed. 1983. Rahway, New Jersey: Merck & Co., Inc.

Minerals Yearbook 1975. 1977. Volume I. Metals, minerals, and fuels.
Bureau of Mines. U.S. Department of the Interior. Washington DC: U.S.
Government Printing Office.

Smith IC, Carson BL, eds. 1980. Trace metals in the environment. Volume
6 - Cobalt. Ann Arbor, Michigan: Ann Arbor Science Publishers, Inc./The
Butterworth Group, 1202 pp.

SELENIUM

MAMMALIAN TOXICITY SUMMARY

I. INTRODUCTION

A. Occurrence and Production

Selenium is widely distributed. It and tellurium are found in small quantities in many sulfide minerals. It is usually produced as a by-product of copper pyrite roasting (USEPA 1980; Beliles 1981).

B. Uses

Selenium is used in photoelectric cells (especially those involved in photocopying), in rectifiers, in glass and ceramics, in pigments, in some metal alloys, and in rubber production. Other uses include as an insecticide and as a topical therapeutic agent for dandruff, acne, etc. (Beliles 1981).

C. Chemical and Physical Properties

Selenium is a nonmetal with properties and reactions similar to those of sulfur. Valences include -2, 0, +4, and +6 (Beliles 1981; USEPA 1980).

II. EXPOSURE AND EXPOSURE LIMITS

A. Oral

Good dietary sources of selenium include seafoods, kidney, liver, and meat (NAS 1980). Plants will take up soluble selenium from the soil, for crop concentrations vary by several orders of magnitude. Estimated United States intake, outside the few, small seleniferous areas, is 132 µg/day (USEPA 1980). In seleniferous areas, the diet may contribute 0.7 to 7 mg Se for a 70-kg man (Snyder et al. 1975). Abuse of selenium supplements may be expected due to the recent publicity for selenium's anticarcinogenic effect (Crounse et al. 1983).

Selenium was found in the finished water of 85 United States cities at levels of < 5 µg/L (USEPA 1975, cited by NAS 1977). Another study of 194 finished water supplies found a mean selenium level of 8 µg/L with a maximum of 10 µg/L (Taylor 1963; cited by NAS 1977).

The national interim primary drinking water standard for selenium is 10 µg/L (USEPA 1975). The ambient water quality criterion for human health is the same (USEPA 1980).

B. Inhalation

Urban regions have particulate selenium concentrations of 0.1 to 10 ng/m^3 (USEPA 1980). Areas around metallurgical industries might have higher concentrations (Synder et al. 1975).

The threshold limit value as a time-weighted average in workroom air for an 8-h day is 0.2 mg/m^3 for selenium and its compounds (ACGIH 1983). The OSHA permissible exposure limit is 0.2 mg/m^3 of Se (OSHA 1981).

C. Dermal

Studies with regular users of selenium sulfide-containing shampoos have found no dermal uptake (NAS 1980).

D. Total Body Burden and Balance Information

According to Snyder et al. (1975), the selenium balance for 70-kg reference man in µg/day is: intake 150 from food and fluids and unknown from air; losses 50 in urine, 20 in feces, and 80 by other routes. Total soft tissue body burden estimate is 13 mg.

III. TOXICOKINETICS

A. Absorption

Selenium compounds are generally well absorbed orally. No data are available on inhalation (Bopp et al. 1982). Lauwerys (1983) states selenium compounds "seem to be easily absorbed through the lungs."

B. Distribution

1. General

Once absorbed, selenium is generally distributed, with highest concentrations in the liver and kidneys, less in the heart, lung, spleen, pancreas and adrenals, and still less in the muscles and brain (Bopp et al. 1982). Normal tissue values are ≦ 1 ppm (Crounse et al. 1983).

2. Blood

Lauwerys (1983) feels that selenium concentrations in blood and urine reflect primarily recent exposure. In the U.S., when the diet contains 62 to 216 µg Se/day, the whole blood concentration is 157 to 265 µg/L. Serum levels are low in persons residing in regions with low-selenium soils. Extreme values are 8 µg/L in the whole blood of persons in the Keshan disease area of China to 3,180 µg/L in an endemic selenosis area on China. Daily intakes at these extremes are 11 and 4,990 µg/day, respectively (Levander 1982). Blood concentrations may not reliably predict clinical toxicity because selenomethionine may be stored in animal proteins and may not be under homeostatic control (Burk 1976). Another complication is a lack of reliability in some reported values. Versieck et al. (1984) cites recent reports

[Alfthan and Kumpulainen 1982; Kumpulainen and Koivistoinen 1981; Ihnat et al. in press] that found erroneous plasma selenium values by even experienced laboratories.

Blood selenium concentrations show homeostatic control once levels of 200 to 240 µg/mL are attained due to chronic ingestion of high selenium in drinking water. Thus, whole blood selenium concentrations were 13.3 to 248 µg/mL for 30 persons drinking contaminated water containing 26 to 1,800 µg Se/L (determined by atomic absorption spectrophotometry with good quality control) (Valentine et al. 1980).

C. Excretion

Excretion is predominantly in urine and is rapid. However, some fecal excretion (via the bile) occurs and significant amounts are exhaled, primarily as dimethyl selenide. At higher doses of selenite, the amount of pulmonary excretion increases, reaching 50% at acutely toxic levels (Bopp et al. 1982). Normal urine values reported range from 7 to 79 µg/L. A biological threshold limit of 100 µg Se/L urine has been proposed based on very limited data (Lauwerys 1983).

IV. EFFECTS

A. Acute and Other Short-Term Exposures

In humans industrially exposed, acute toxicity is primarily due to the irritative and allergenic properties of selenious acid (H_2SeO_3), formed from water and selenium dioxide. SeO_2 is formed whenever selenium is heated in air. Symptoms are nonspecific--redness of mucous membranes and eyes, sneezing, coughing, perhaps dyspnea, and frontal headache (Beliles 1981).

Livestock in seleniferous areas may get "blind staggers," which is characterized by some impairment of vision, a difficulty in judging near objects, and a tendency to wander. Severe cases have abdominal pain, paralysis, and death from respiratory failure (USEPA 1980).

B. Chronic Exposure

Livestock may develop "alkali disease," chronic selenium poisoning. Effects seen include emaciation, loss of hair from mane and tail, separation of the hoof, atrophy and decompensation of the heart, liver cirrhosis, renal glomerulonephritis, and anemia (USEPA 1980).

Chronic poisoning occurs in livestock or laboratory animals consuming diets containing > 2 ppm in drinking water or > 3 ppm in food. Where selenium excess or deficiency in soils produces disease in livestock, NAS (1980) states that no disease states attributable to the excess or deficiency have been reported in humans. However, Snyder et al. (1975) cite Rosenfeld and Beath (1964) as reporting that chronic toxicity resulted from humans ingesting 1 mg Se/kg/day. Human effects that correlated somewhat with selenium concentrations in the urine were reported in the 1930's, e.g., gastrointestinal complaints, jaundice, skin hyperpigmentation, nail changes, bad teeth, arthritis, dizziness, and fatigue. A higher incidence of dental caries has been reported for children growing up in geographic areas with high-selenium soils (Crounse et al. 1983).

C. Biochemistry

1. Effects on Enzymes

Selenium's best known biochemical function is as a component of the enzyme glutathione peroxidase. The enzyme prevents oxidative damage of important cell constituents (NAS 1980).

2. Metabolism

Extensive metabolism of selenium takes place, principally in the liver. Especially noteworthy is the reduction of selenite ($SeO_3{}^2$) or selenate ($SeO_4{}^2$) to hydrogen selenide (H_2S), which is then methylated to dimethyl selenide and exhaled or to the trimethylselenonium ion and excreted in the urine (Bopp et al. 1982). Selenomethionine will be incorporated into tissue protein in place of methionine (Burk 1976).

3. Antagonisms and Synergisms

There is a direct, mutual antagonism between selenium and arsenic toxicity. Similar effects have been seen with cadmium, mercury, silver, and thallium (USEPA 1980). The actions of selenium and vitamin E are synergistic (Crounse et al. 1983).

4. Physiological Requirements

Selenium is an essential nutrient. NAS (1980) recommends 50 to 200 µg/day as adequate and safe for adults. In animals, selenium deficiencies occur when diets contain < 0.02 to 0.05 ppm Se. If diets are also deficient in vitamin E, more selenium is required.

An endemic cardiomyopathy probably due to ultra-low selenium deficiency in soils of Keshan province in China was reported in 1980. In Keshan province, average human blood levels were < 0.01 ppm compared to an average 0.068 ppm in human blood from low-selenium areas in New Zealand and Egypt. Selenium supplementation prevented or ameliorated the syndrome (Crounse et al. 1983).

D. Specific Organs and Systems

Other than the symptoms listed above, no organ-specific effects have been noted.

E. Teratogenicity

Selenium is teratogenic, especially to chicks and sheep (USEPA 1980). There is some limited evidence that it is embryotoxic and perhaps teratogenic in man (USEPA 1980; Beliles 1981).

F. Mutagenicity

Selenium compounds have been found mutagenic to barley and nonmutagenic to human leukocyte and fibroblast cultures (USEPA 1980).

G. Carcinogenicity

Selenium monosulfide, the ingredient of Selsun® antidandruff shampoo, was carcinogenic in rats and female mice by gavage, producing malignancies in liver and lung (NCI 1980a) but not by dermal application (NCI 1980b). The picture is complicated by several reports that the presence of low environmental levels of selenium is anticarcinogenic (USEPA 1980).

V. REFERENCES

ACGIH. 1983. TLVs® Threshold limit values for chemical substances and physical agents in the work environment with intended changes for 1983-84. Cincinnati, Ohio: American Conference of Governmental Industrial Hygienists.

Beliles, RP. 1981. Chapter 31. Phosphorus, selenium and tellurium. In: Patty's industrial hygiene and toxicology. 3rd ed. Volume IIA, Toxicology. Clayton GD, Clayton FE, eds. New York: A Wiley-Interscience Publication. John Wiley and Sons, pp. 2121-2140.

Bopp BA, Sonders RC, Kesterson JW. 1982. Metabolic fate of selected selenium compounds in laboratory animals and man. Drug Metab Rev 13(3):271-318.

Burk, RF. 1976. The significance of selenium levels in blood. In: Proc. symp. on selenium-tellurium in the environ. Pittsburgh, PA: Industrial Health Foundation, Inc., pp. 194-201.

Crounse RG, Pories WJ, Bray JT, Mauger RL. 1983. Geochemistry and man: Health and disease 1. Essential elements. In: Applied environmental geochemistry. Thornton, I., ed., London, UK: Academic Press, pp. 309-333.

Lauwerys RR. 1983. Chapter II. Biological monitoring of exposure to inorganic and organometallic substances. In: Industrial chemical exposure: Guidelines for biological monitoring. Davis, CA: Biomedical Publications, pp. 9-50.

Levander OA. 1982. Clinical consequences of low selenium intake and its relationship to vitamin E. Ann NY Acad Sci 393:70-82.

NAS. 1977. Natl. Academy of Sciences. Drinking water and health. Vol. I. Washington, DC.

NAS. 1980. Natl. Academy of Sciences. Recommended dietary allowances. 9th ed. Washington, DC: Printing and Publishing Office, National Academy of Sciences.

NCI. 1980a. National Cancer Institute. Bioassay of selenium sulfide
(gavage) for possible carcinogenicity. Springfield, VA: National Technical
Information Service. NCI-CG-TR-194, NIH/PUB-80-1750, NTP-80-17, PB82-164955.

NCI. 1980b. National Cancer Institute. Bioassay of selenium sulfide
(dermal study) for possible carcinogenicity. Springfield, VA: National
Technical Information Service. NCI-CG-TR-197, NIH/PUB-80-1753, NTP-80-18,
PB82-165291.

OSHA. 1981. Occupational Safety and Health Admin. Occupational safety and
health standards. Subpart 2--Toxic and hazardous substances. Code of
federal regulations 29 (Part 1910.1000) pp. 673679.

Snyder WS, Cook MJ, Nasset ES, Karhausen LR, Howells GP, Tipton IH. 1975.
International Commission on Radiological Protection. Report of the task
group on reference man. New York. ICRP Publication 23.

USEPA. 1975. U.S. Environmental Protection Agency. National interim
primary drinking water regulations. (40 FR 59566).

USEPA. 1980. Ambient water quality criteria for selenium. Springfield,
Virginia: National Technical Information Service. PB81-117814.

Valentine JL, Kang HK, Dang P-M, Schluchter M. 1980. Selenium
concentrations and glutathione peroxidase activities in a population exposed
to selenium via drinking water. J Toxicol Environ Health 6:731-736.

Versieck J. 1984. Trace element analysis - a plea for accuracy. Trace
Elements Med 1(1):2-12.

SILVER

MAMMALIAN TOXICITY SUMMARY

Comprehensive reviews of silver toxicity may be found in USEPA (1980), Stokinger (1981), Smith and Carson (1977), and Nordberg and Fowler (1977). The ambient water quality criteria document for silver (USEPA 1980) contains a 169-page section on the human health effects of silver, which was chiefly written by the author of the present profile.

The major problem in humans arising from over exposure to silver is called argyria. Generalized argyria is characterized by a slate blue-gray coloration of the skin, mucous membranes, and internal organs (Smith and Carson 1977).

I. INTRODUCTION

A. Occurrence and Production

The average crustal abundance of silver is ~ 0.1 ppm. Silver occurs in nature as native silver, natural alloys, and various minerals, especially silver-bearing galena (PbS). In the United States silver is recovered from Ag-< 5% Pb, Au, Pb, Cu-Pb-Zn, Pb-Ag, and Zn ores. Comparable amounts of silver are recovered annually from photographic wastes and other secondary sources (Smith and Carson 1977).

B. Uses

Major uses (≥ ~ 10% each) for silver include electroplated ware, sterling ware, photographic materials, brazing alloys and solder, and electrical contacts and conductors. Minor uses (~ 5% or less) include jewelry, dental and medical supplies, batteries, catalysts, and bearings. The widespread medical use of silver compounds declined about 40 years ago. because of the advent of the sulfa drugs and antibiotic antimicrobials as well as the fear of developing the cosmetic disfigurement argyria. The use of silver in the oligodynamic purification of drinking and bottling waters and of swimming pools has undergone a renascence in the United States in the last decade. Cloud seeding with AgI is a minor use (Smith and Carson 1977).

C. Physical and Chemical Properties

Silver, atomic no. 47, melts at 960.8°C and boils at 2212°C. It has the highest thermal and electrical conductivity of all the metals. Its compounds exhibit +1, +2, and +3 oxidation states; the +1 state is the more common and the only one stable in aqueous media in the environment. Unlike the metals of the IA group, "AgOH," Ag_2S, and Ag halides except AgF are water insoluble (Smith and Carson 1977).

II. EXPOSURE AND EXPOSURE LIMITS

A. Oral

The USEPA interim drinking water standard and 1980 criterion for silver in drinking water is 50 µg/L. Scarcely any natural waters contain near that amount (USEPA 1980). Silver levels in the finished water of the 100 largest U.S. cities were as high as 7 µg/L with a median of 2.3 µg/L (Durfor and Becker 1964; cited by NAS 1977). Another study of 380 samples of finished-water found silver in 6.1% with concentrations from 0.3 to 5 µg/L, the mean was 2.2 µg/L (Kopp 1973; cited in NAS 1977). Another sampling of tap water found none of 2,595 samples exceeded the interim drinking water standard, the maximum level was 30 µg/L (McCabe 1970; cited by NAS 1977). From the Chemical Analysis of Interstate Carrier Water Supply Systems, silver was found at the < 0.1 µg/L level in 45% of the analyses, and 99.5% were less than 50 µg/L (USEPA 1975).

The original 1962 U.S. Public Health Standard was meant to prevent adding silver to water for its purifying effect at higher levels. Silver at 100 to 200 µg/L was used on the Apollo spacecraft and on Soviet spaceships to purify drinking water and wastewater. Numerous silver medications were widely used 40 or more years ago. As recently as the 1970's, a case of argyria was reported in a middle-aged woman who treated her irritated gums with numerous $AgNO_3$ astringent sticks ("lunar caustic") within a few years' time. Silver amalgam fillings also appear to contribute to the body burden of silver (USEPA 1980).

B. Inhalation

Reported average ambient air concentrations range from 0.18 ng/m³ in Chadron, Nebraska, to 10.5 ng/m³ in Kellogg, Idaho, near where silver-lead ores were mined and smelted (USEPA 1980).

The threshold limit value as a time-weighted average in workroom air for an 8-h day is 0.1 mg Ag/m³ for silver metal (ACGIH 1983). The OSHA permissible exposure limit is 0.01 mg/m³ for silver metal and soluble compounds (OSHA 1981).

C. Dermal

A few drops of silver nitrate (or antibiotics) are applied to the conjunctiva of newborn infants to prevent ophthalmia neonatorium due to gonorrhea transmitted from the mother. Silver salts in ointment form, silver pyridazine, colloidal silver isotonic solutions, and hypotonic $AgNO_3$ solutions have been used in burn wound treatment. The plasma of patients treated for up to 80 days with silver preparations had up to 0.30 mg Ag/L. In the past, argyrosis of the conjunctiva sometimes developed from the use of $AgNO_3$ hair dyes to dye the eyebrows and eyelashes. Use of silver in swimming pools allows dermal contact, but absorption by intact skin is unlikely (USEPA 1980).

D. Total Body Burden and Balance Information

Older and more modern determinations of silver in food fall within the same order of magnitude, all giving somewhat less than 100 µg/day in the diet.

According to Snyder et al. (1975) the silver balance for 70-kg reference man (µg/day) is as follows:

Intake, µg/day		Losses, µg/day	
Food and fluids	70	Urine	9
		Feces	60
		Sweat	0.4
		Hair	0.6

III. TOXICOKINETICS

A. Absorption

Occupational argyria has generally been caused by absorption from the lungs or the conjunctiva. Even nonintact skin appears to absorb < 1% of an applied dose. When given in single oral doses, up to 10% of the silver is absorbed. Most of an accidentally inhaled dose of ^{110m}Ag by a human had a biological half-life of one day, probably due to mucociliary clearance, swallowing, and fecal excretion. Possibly colloidal forms were absorbed in the lungs. Phagocytosis may have accounted for the activity seen in the liver on the second day (USEPA 1980).

B. Distribution

1. General

In persons who had never used silver medications, the primary deposition sites were the liver, skin, adrenals, lungs, muscle, pancreas, kidneys, heart, and spleen. In argyria, silver was deposited in the blood vessels and connective tissue, especially around the face, conjunctiva, hands, and fingernails. Aside from these sites, the dermis of the skin, glomeruli of the kidney, choroid plexus, mesenteric glands, and thyroid had the highest silver concentrations. In radiotracer experiments with rats, the more carrier, the more silver was found in the liver and skin (USEPA 1980).

2. Blood

Whole blood of normals contained 24 µg Ag/g (determined by emission spectroscopy) whereas serum from 122 blood donors contained 0.004 µg/g. Hospital workers had 0.013 to 0.014 µg/g in their serum (7,148 below the detection limit), but the patients had 0.03 µg/g (Stokinger 1981).

Silver is carried in the protein fractions of plasma as a silver-protein complex expecially with the globulins. (Silver's tissue deposition apparently indicates vascular permeability to protein.) Blood clearance is rapid. Rabbits injected intraveneously with radiosilver showed 86% of the injected dose already in the liver within 5 min. In a patient dying of a malignant carcinoid, an injection of 43 µCi radiosilver was rapidly removed from the blood to the liver. Seven minutes after the injection, only 30% of the injected dose was present in the blood (USEPA 1980).

C. Excretion

Excretion of absorbed silver is predominantly fecal via the bile. The deposition of silver in the skin and connective tissues in argyria appears to be an excretory pathway. The deposits are so inert that they cannot be removed by common (or uncommon) heavy metal detoxification procedures. Silver is only detected in human urine in poisoning cases although it can be detected in normal human blood (USEPA 1980).

IV. EFFECTS

A. Acute and Other Short-Term Exposures

Large oral doses of silver nitrate, which is a corrosive, have been taken by persons attempting suicide. The patient suffers violent abdominal pain, abdominal rigidity, vomiting, convulsions, and severe shock. The gastroenteritis is sometimes fatal. Necrosis and hemorrhages of the bone marrow, liver, and kidney were observed in patients dying from intravenous doses of Collargol® (silver plus silver oxide) (USEPA 1980).

Toxic effects (e.g., electrolyte disturbances) that are only indirectly attributable to silver have occurred in burn patients treated topically with silver compounds. Most effects develop within a few days of the initial application and can be corrected by appropriate medical treatment (USEPA 1980).

Intravenous dosing is the major route for animal acute toxicity. Silver nitrate injections (e.g., ~ 32 mg Ag/kg in dogs) are used to produce acute pulmonary edema for study. Inorganic silver compounds given in lethal intravenous doses produce weakness, rigidity, and contractures in the legs and loss of voluntary movements and interfere with blood supply to the heart (USEPA 1980).

B. Chronic Exposure

Generalized argyria, a slate gray pigmentation of the skin, hair, and internal organs caused by silver deposition; localized argyria; and argyrosis (usually used for the rare condition argyria of the eye) are the most noticeable effects of chronic* human exposure to silver.

---

*   Rarely, argyria has resulted from acute exposure. The most sudden onset was after 2 days' treatment with Protargol® by urethral application (USEPA 1980).

Every widely used silver compound has caused argyria, usually due to absorption by exposed mucous membranes. Medical treatment with silver compounds was the most common cause. The occupational disease was not common and occurred almost exclusively among silver nitrate makers. The bronchitis and emphysema described in silver workers has not been successfully attributed to silver exposure (USEPA 1980).

Numerous chronic oral animal studies of questionable quality are reviewed by USEPA (1980). Levels of 50 µg/L in the drinking water of rats for 5 months showed no effects. At $\geq$ 400 µg/L, the kidneys showed hemorrhages and at 500 µg/L or 2,000 µg/L, conditioned reflexes and immunological resistance were lowered. A level of 20 mg/L produced numerous physiological changes including growth depression.

1. Effects on enzymes

Silver interferes with the activity of glutathione peroxidase (a selenium-containing enzyme) and alters the metabolism of cyclic adenosine monophosphate (AMP) (USEPA 1980).

2. Metabolism

Silver in the blood appears to be phagocytized by the reticuloendothelial system and removed by the liver. Silver selenide may form in the liver and silver-metallothionein complex may form in the kidney. The nature of the inert deposits in the skin is most probably metallic silver, silver sulfide, or silver complexes with the sulfhydryl amino acids in proteins (USEPA 1980).

3. Antagonisms and synergisms

Silver is precipitated by protein and chloride ion. Thus, NaCl (table salt) is an antidote for $AgNO_3$ poisoning. Silver antagonizes selenium, vitamin E, and copper and will induce deficiency symptoms in animals fed adequate diets or aggravate deficiency symptoms in animals whose diets lack one or more of the nutrients. Reports have described these effects in rats, dogs, sheep, pigs, chicks, turkey poults, and ducklings. Chicks develop exudative diathesis. Liver necrosis, growth depression, cardiac enlargement, muscular dystrophy, and mortality are among the deficiency signs and symptoms (USEPA 1980).

D. Specific Organs and Systems

Several organs may be affected by exposure to silver sufficient to cause copper, selenium, and/or vitamin E deficiency. These changes are not discussed here.

1. Skin, eye, and connective tissue of internal organs

Silver is deposited in these tissues leading to discoloration, which is darkest in areas exposed to light (USEPA 1980).

A few cases of dermatitis and hypersensitivity have been reported (USEPA 1980).

2. Lung

Lethal injections of silver nitrate in laboratory mammals are used as a model of pulmonary lung edema.

Bronchitis and emphysema in chronically exposed silver workers were not conclusively attributed to silver exposure (USEPA 1980).

3. Gastrointestinal tract

Large acute doses of silver nitrate produce gastroenteritis, which may be fatal (USEPA 1980).

E. Teratogenicity

No evidence of silver teratogenicity has appeared in the literature (USEPA 1980).

F. Mutagenicity

Silver compounds were not mutagenic in tests with Escherichia coli, Micrococcus species, Bacillus subtilis, and Salmonella typhimurium (the Ames test). They did cause chromosome breakage in germinated pea (Pisum) seeds (USEPA 1980).

G. Carcinogenicity

Implants of silver foils or disks or injections of metallic silver in colloidal suspension have produced tumors or hyperplasia; but the effects were usually attributed to the particular physical form of the metal, to its being an exogenous irritant, or to its lowering resistance due to some solubilized Ag^+. None of these studies appears to be sufficient as evidence that silver or its compounds are carcinogens (USEPA 1980).

V. REFERENCES

ACGIH. 1983. TLVs[®] Threshold limit values for chemical substances and physical agents in the work environment with intended changes for 1983-84. Cincinnati, Ohio: American Conference of Governmental Industrial Hygienists.

NAS. 1977. National Academy of Sciences. Drinking water and health. Washington, DC: Printing and Publishing Office, National Academy of Sciences.

Nordberg GF, Fowler BA. 1977. Silver. In: Toxicology of metals-Volume II. Springfield, Virginia. National Technical Information Service, pp. 358-366. PB-268-324.

OSHA.   1981. Occupational Health and Safety Administration.   Occupational safety and health standards.   Subpart 2--Toxic and hazardous substances. Code of federal regulations 29 (Part 1910.1000): 673-679.

Smith IC, Carson BL.   1977.   Trace metals in the environment.   Volume 2--Silver. Ann Arbor, Michigan: Ann Arbor Science Publishers, Inc., 469 pp.

Stokinger HE.   1981.   Chapter 29.   The metals.   In: Patty's industrial hygiene and toxicology, 3rd ed., volume IIA, Toxicology. Clayton DG, Clayton FE, eds. New York: Wiley-Interscience, John Wiley & Sons, pp. 1493-2060.

USEPA.   1975.   Environmental Protection Agency National Interim Primary Drinking Water Regulations.   40CFR 141 as corrected through 47FR 10998, March 12, 1982.

USEPA.   1980.   U.S. Environmental Protection Agency.   Ambient water quality criteria for silver.   Springfield, VA: National Technical Information Service.   PB81-117822.

SODIUM

MAMMALIAN TOXICITY SUMMARY

I. INTRODUCTION

A. Occurrence and Production

In the United States, soda ash (technical grade Na_2CO_3) is the major sodium compound produced. Natural soda ash is produced by electrolysis of sea water or lake brines. Natural sodium sulfate also occurs in brines. Synthetic soda ash is produced by the ammonia soda or Solvay process. Sodium chloride (table salt) is produced by mining rock salt, evaporating underground salt deposit brines, or evaporating sea water in the sun. NaOH (caustic soda or soda lye) may be produced by treating $Ca(OH)_2$ with NaCl or by electrolyzing NaCl. Sodium hypochlorite (NaOCl) is produced from Cl_2 and NaOH; sodium chlorite ($NaClO_2$), from ClO_2 and NaOH. Sodium metal is produced by electrolytic reduction (Merck Index 1983).

B. Uses

Sodium compounds are among the highest tonnage industrial chemicals. Major industrial uses of sodium compounds include manufacturing sodium glasses and detergents; bleaching pulp, paper, and textiles; and treating water. Sodium metal is used to manufacture sodium compounds and tetraethyllead. It is used in organic syntheses and in sodium lamps and photoelectric cells. Household bleach contains 5% NaOCl, and household water softening devices replace hardwater ions (calcium and magnesium) with Na^+. Sodium bicarbonate is used in baking, beverages, and antacids (Merck Index 1983).

C. Chemical and Physical Properties

Sodium, atomic no. 11, is the most familiar alkali metal. It is soft at ordinary temperatures and so reactive with water and oxygen it is stored under oxygen-free liquids such as kerosene. It melts at 97.82°C and boils at 881.4°C. Practically all of its compounds are water soluble. Generally the chemistry of the anion dominates the chemical behavior and industrial use. The pH of a 5% solution of NaOH is ∿ 14 (Merck Index 1983).

II. EXPOSURE AND EXPOSURE LIMITS

A. Oral

Good natural sources of dietary sodium are cheese, milk, and shellfish. Meat, fish, poultry, and eggs contribute somewhat less. Most dietary sodium has been added to the prepared food or is added at the table as salt (NaCl). At 2 mo, infants consume ∿ 300 mg/day; at 12 mo, ∿ 1,400 mg/day. (Human milk contains 161 mg Na/L and cow's milk contains ∿ 483 mg/L.) Adults need little more but will consume 2,300 to 6,900 mg/day when they have free access to salt (NAS 1980).

In a study of 2,100 U.S. water supplies, the level of sodium in the finished water ranged from 0.4 to 1,900 mg/L with 42% in excess of 20 mg/L and 4% greater than 250 mg/L (White et al. 1967; cited by NAS 1977). From the Chemical Analysis of Interstate Carrier Water Supply Systems, sodium concentrations in 630 systems ranged from < 1 to 402 mg/L with 42% greater than 20 mg/L and 3% over 200 mg/L (NAS 1977). Carbonated beverages contribute significant amounts of sodium to the daily intake.

B. Inhalation

The major source of sodium in the air is sea salt (brines).

C. Dermal

A direct source of sodium to the skin would be softened water and sodium bicarbonate topical applications for dermal irritation.

D. Total Body Burden and Balance Information

Snyder et al. (1975) summarized sodium balance information for 70-kg reference man (total body burden 100 g Na):

Intake, mg/day	Losses, mg/day	
Food and fluids 4,400	Urine	3,300
	Feces	100
	Sweat	870
	Other fluids	130
	Hair	0.1

Hard physical labor at high ambient temperatures may cause sodium losses in the sweat as high as 8,000 mg/day. If sweat losses require drinking > 3 L/day, NaCl supplements of 2,000 to 7,000 mg Na/day/L of extra water loss are recommended (NAS 1980).

III. TOXICOKINETICS

A. Absorption

Gastrointestinal absorption of sodium in the diet is 100% (NAS 1980).

B. Distribution

1. General

Sodium concentrations in tissues and body fluids are homeostatically controlled. The sodium content generally determines the volume of extracellular fluid retained by the tissues; this volume may become distorted to maintain otherwise normal sodium concentrations (NAS 1980).

2. Blood

Whole blood of 70-kg reference man contains 10 g sodium, most of which is in the plasma (Snyder et al. 1975). Normal average serum sodium values from colorimetric determinations are 2,900 to 3,560 mg/L (Henry 1964). Plasma protein concentrations decrease during increases of extracellular fluid volume and may indicate volume disturbance due to excess sodium intake. During vigorous exercise, sodium in the plasma increases (NAS 1980). There is a slight increase after meals, but there is no significant sex difference (Henry 1964).

3. Adipose

Snyder et al. (1975) give 7.6 g ($\sim$ 510 µg/g) as the total adipose Na content for reference man.

C. Excretion

The kidneys excrete sodium in the urine. The action of the hormone aldosterone on renal tubular functions maintains sodium homeostasis. In elevated sodium intakes, aldosterone levels decrease allowing increased sodium excretion in the urine (NAS 1980).

IV.   EFFECTS

A. Acute and Other Short-Term Exposures

The acute toxicity of sodium compounds is due either to their high pH (e.g., NaOH) and concomitant corrosivity to mucous membranes and skin or to the toxicity of their anions. Neither will be discussed in this brief review.

High doses of NaCl are used therapeutically as an emetic (Merck Index 1983). Early symptoms of mass NaCl poisonings in infants (due to accidental substitution of salt for cane sugar) include nausea, vomiting, and refusal of feeding. Central nervous system effects develop: convulsions or muscular twitching. Body temperatures may rise. Fast (tachypnea) and difficult breathing (dyspnea) indicate severe respiratory distress. Temporary redistribution of water affects the brain. In infants who have died, extensive hemorrhagic phenomenon were observed in the central nervous system and kidney tubules. Peritoneal dialysis is the treatment of choice for salt-poisoned infants (Arena 1973).

B. Chronic Exposure

Salt poisoning in infants, described in IV.A, may occur in newborns with continuous use of skim milk, which is high in natural NaCl (Arena 1973).

Epidemiological studies have indicated that long-term excessive sodium intake is one of many factors associated with hypertension in humans. A high Na:K ratio in the diet may be detrimental to persons susceptible to high blood pressure.   Some adults, however, tolerate chronic intakes > 40 g NaCl/day (NAS 1980).

Feeding high salt diets to laboratory animals leads to hypertension, especially if salt-sensitive species are started on the regime early in infancy (NAS 1980).

C.   Biochemistry

   1.   Effects on Enzymes

No information.

   2.   Metabolism

The primary function of sodium ions is the maintenance of osmotic equilibrium and extracellular fluid volume (NAS 1980).   No information on metabolic changes of sodium was found.

   3.   Antagonisms and Synergisms

No information.

   4.   Physiological Requirements

Sodium requirements for infants and children under 10 range from 115 to 1,800 mg/day, depending on age.   For adolescents ≥ 11 yr, the requirement is 900 to 2,700 mg/day; for adults, 1,100 to 3,300 mg/day.

Inadequate replacement of salt during sweating will lead to salt depletion heat exhaustion, characterized by fatigue, nausea, giddiness, vomiting, and exhaustion.   Fainting may occur with only moderate depletion. Heat stroke may occur suddenly upon loss of consciousness.   Hot dry skin, high body temperature, and signs of cerebral dysfunction are characterstic. Free access to water must be allowed when salt tablets are given those exercising vigorously at high ambient temperatures to prevent hypernatremia (NAS 1980).

D.   Specific Organs and Systems

   1.   Gastrointestinal Tract

Excess acute sodium intake leads to nausea and vomiting (Arena 1973).

   2.   Central Nervous System

Convulsions or muscular twitching and extensive hemorrhaging in the central nervous system are observed in infants acutely poisoned by NaCl (Arena 1973).

3. Kidneys

Excessive acute sodium intake damages the kidney tubules (Arena 1973).

E. Teratogenicity

Injections of 1,900 or 2,500 mg NaCl/kg into mice on day 10 or 11 of pregnancy gave $\leq$ 18% skeletal defects, especially clubfoot (Nishimura and Miyamoto 1969; cited by Shepard 1980).

F. Mutagenicity

No information.

G. Carcinogenicity

No information.

V. REFERENCES

Arena JM. 1973. Poisoning: toxicology--symptoms--treatments. Springfield, IL: Charles C. Thomas Publisher.

Henry RJ. 1964. Clinical chemistry: principles and technics. New York, NY: Hoeber Medical Division, Harper & Row Publishers.

Merck Index, 10th ed. 1983. Rahway, New Jersey: Merck & Co., Inc.

NAS. 1977. National Academy of Sciences. Drinking water and health. Vol. I. Washington, DC.

NAS. 1980. National Academy of Sciences. Recommended dietary allowances, 9th ed. Washington, DC.

Shepard TH. 1980. Catalog of teratogenic agents, 3rd ed. Baltimore: The Johns Hopkins University Press.

Snyder WS, Cook MJ, Nasset ES, Karhausen LR, Howells GP, Tipton IH. 1975. International Commission on Radiological Protection. Report of the task group on reference man. New York. ICRP Publication 23.

STRONTIUM

MAMMALIAN TOXICITY SUMMARY

I. INTRODUCTION

A. Occurrence and Production

Strontium is widely found, generally in association with calcium. Seawater contains 10 ppm. The commerical minerals are celestite ($SrSO_4$) and strontianite ($SrCO_3$). The metal is produced by electrolysis or by thermal reduction with aluminum (Stokinger 1981).

B. Uses

The major use is for the red color of pyrotechnics; strontium nitrate is the usual compound. Other compounds are used in ceramics, corrosion inhibitors, depilatories, luminous paints, etc. (Stokinger 1981).

C. Chemical and Physical Properties

Strontium is a typical divalent alkaline earth. Compounds are typically a bit more soluble than those of barium (Stokinger 1981).

II. EXPOSURE AND EXPOSURE LIMITS

A. Oral

Estimates of dietary Sr intake range from 0.98 to 2.2 mg Sr/day for adults, about one-third of which is from milk (Snyder et al. 1975).

B. Inhalation

No data.

C. Dermal

No data.

D. Total Body Burden and Balance Information

According to Snyder et al. (1975), the strontium balance for 70-kg reference man in mg/day is: intake from food and fluids 1.9; losses of 0.34 in urine, 1.5 in feces, 0.02 in sweat, 0.0002 in hair, and a trace in other fluids.

It is estimated that the average man contains 320 mg of strontium, 99% in the bones (Stokinger 1981; Snyder et al. 1975).

III. TOXICOKINETICS

A. Absorption

Absorption from the gastrointestinal tract is poor, but is increased during calcium-poor diets (Stokinger 1981). Estimates of gastrointestinal absorption of ^{90}Sr ranged from 9 to 63% ($\sim$ 38% av.) (Snyder et al. 1975).

B. Distribution

1. General

The skeleton contains most absorbed strontium (Stokinger 1981).

2. Blood

Snyder et al. (1975) give 180 µg Sr in whole blood of reference man, with 170 µg in plasma and 0.008 µg in erythrocytes. Average Sr concentrations in the plasma of eight normal adults was 29 µg/L (16 to 43 µg/L) (Stokinger 1981).

3. Adipose

Snyder et al. (1975) estimate 370 µg (25 µg/g) as the strontium content of adipose tissue.

C. Excretion

Excretion of ingested strontium is primarily in the feces (Stokinger 1981).

IV. EFFECTS

A. Acute and Other Short-Term Exposures

Acute strontium toxicity causes death by respiratory failure. Most of its compounds have a low order of toxicity even by intravenous injection (Stokinger 1981).

B. Chronic Exposure

Evidence for chronic effects in animals presented in Stokinger (1981) is negligible and hard to interpret. The tendency appears to be disturbances of mineral metabolism.

C. Biochemistry

1. Effects on Enzymes

Strontium substitutes for calcium in many normal mechanisms often with no apparent ill effect (Stokinger 1981).

2. Metabolism

Strontium is not metabolized.

3. Antagonisms and Synergisms

Interactions with calcium are probably responsible for all of strontium's effects.

D. Specific Organs and Systems

Stokinger (1981) cites functional and histological changes in several organs and systems that were interpreted as due to interference with calcium.

E. Teratogenicity

Shepard (1980) lists skeletal defects of dogs and mice due to ^{90}Sr. No information was found on stable strontium.

F. Mutagenicity

No data.

G. Carcinogenicity

No data.

V. REFERENCES

Shepard TH. 1980. Catalog of teratogenic agents, 3rd ed. Baltimore: The Johns Hopkins University Press.

Snyder WS, Cook MJ, Nasset ES, Karhausen LR, Howells GP, Tipton IH. 1975. International Commission on Radiological Protection. Report of the task group on reference man. New York. ICRP Publication 23.

Stokinger HE. 1981. Chapter 29. The metals. In: Patty's industrial hygiene and toxicology. 3rd ed. Volume 2A, Toxicology. Clayton GD, Clayton FE, eds. New York: A Wiley-Interscience Publication. John Wiley and Sons, pp. 1493-2060.

TANTALUM

MAMMALIAN TOXICITY SUMMARY

I. INTRODUCTION

A. Occurrence and Production

Tantalum is one of the rarer elements in the earth's crust, ranking 54th. It is almost always found associated with columbium in minerals such as ferrous manganese tantalate niobate $(Fe,Mn)(Ta,Nb)_2O_6$ called tantalite or columbite depending on predominating metal. It is also found in euxenite $(Y,Ce,Cs,U,Th)(Nb,Ta,Ti)_2O_6$ and microlite $(Na,Ca)_2Ta_2(O,OH,F)$ (Stokinger 1981; Browning 1969).

B. Uses

Tantalum is used in the manufacture of capacitors and other electronic components, corrosion resistant chemical equipment, and steel alloys. TaC is used in cutting tools (Minerals Yearbook 1977). Inert tantalum powder is used widely in surgical repairs and diagnostic tests (Stokinger 1981).

C. Chemical and Physical Properties

Tantalum, atomic numer 73, melts at 3000°C, boils at > 4100°C. It is insoluble in hot or cold water and acids and soluble in hydrogen fluoride and fused alkalies. It exhibits oxidation states of +2 to +5, but only +5 forms are stable (Stokinger 1981; Browning 1969).

II. EXPOSURE AND EXPOSURE LIMITS

A. Oral

No information.

B. Inhalation

The threshold limit value as a time-weighted average in workroom air for an 8-h day is 5 mg/m³ for tantalum; the short-term exposure limit is 10 mg/m³/15 min (ACGIH 1983). The OSHA permissible exposure limit is 5 mg/m³ (OSHA 1981).

C. Dermal

No information.

D.   Total Body Burden and Balance Information

No information.

III. TOXICOKINETICS

A.   Absorption

Little has been reported about the absorption of tantalum. When given oral doses of radioactive tantalum, young suckling rats absorbed amounts several times those of the adults, but they also experienced rapid loss of the compound by weaning time (Shiraishi and Ichikawa 1972; cited by Stokinger 1981). In rats fed the soluble potassium tantalate, < 1% was retained in the body after 14 days (Fleshman 1971; cited by Stokinger 1981).

B.   Distribution

1.   General

In the Fleshman feeding study mentioned above, 40% of the absorbed tantalum was deposited in the bone. In workers exposed to radioactive tantalum accidently, the area of maximum radioactivity moved from the upper respiratory area the first day to the lower intestinal tract by the third day and remained there (Sill et al. 1969; cited by Stokinger 1981.

2.   Blood

By spark source mass spectrometry, a single normal blood sample was found to have 16 µg/kg compared to 5 and 240 µg/L for Ta in two normal urine samples (Stokinger 1981).

C.   Excretion

Ingested tantalum appears to be excreted from the gastrointestinal tract, 93% in 7 days in the Sill study mentioned above. With inhaled tantalum, there appears to be early, rapid tracheobronchial clearance with prolonged (> 12 mo) time required for alveolar clearance (Stokinger 1981).

IV. EFFECTS

A.   Acute and Other Short-Term Exposures

Tantalum is relatively nontoxic; in fact, tantalum oxide appears to be nontoxic: oral doses as high as 8000 mg/kg caused no effect in rats. The soluble tantalum pentachloride had an oral LD_{50} of 958 Ta/kg and an intraperitoneal LD_{50} of 38 mg Ta/kg (Cochran et al. 1950).

B.   Chronic Exposure

Tantalum appears to be very nontoxic in chronic studies. A study of workers exposed to tantalum carbide dust in the hard-metal cutting tool industry combined with a histological study of the lungs of rats similarly exposed found the tantalum carbide dust acted as a physiologically inert substance (Miller et al. 1953; cited by Stokinger 1981). Early pulmonary fibrosis was observed in Russian chemical workers and welders exposed to both Ta and Nb (Stokinger 1981). In another study, guinea pigs were exposed to tantalum oxide at intratracheal doses of 100 mg/animal. Initially, there were acute responses including bronchial epithelial hyperplasia and hypertrophic focal emphysema; but after 1 yr, almost complete recovery had occurred (Schepers 1955; cited by Stokinger 1981).

In addition to these studies, the fact that tantalum metal products such as wire and gauge and tantalum powder have been used in surgery and other clinical applications for > 25 yr with no reported adverse effect demonstrates tantalum's nontoxicity (Stokinger 1981).

C.   Biochemistry

1.   Effects on Enzymes

No information.

2.   Metabolism

No information.

3.   Antagonisms and Synergisms

No information.

D.   Specific Organs and Systems

Tantalum and many of its compounds appear to be remarkably physiologically inert. Lung changes seen in animals inhaling Ta oxide were not fibrogenic and were almost completely reversible (Stokinger 1981).

E.   Teratogenicity

No information.

F.   Mutagenciity

No information.

G.   Carcinogenicity

One study found malignant fibrosarcomas in rats that had tantalum metal foil embedded for 714 days (Oppenheimer et al. 1957; cited by Stokinger

1981). This contrasts with > 25 yr of surgical use in humans with no reported carcinogenic response.

V. REFERENCES

ACGIH. 1983. TLVs[®] Threshold limit values for chemical substances and physical agents in the work environment with intended changes for 1983-84. Cincinnati, Ohio: American Conference of Governmental Industrial Hygienists.

Browning E. 1969. Toxicity of industrial metals. 2nd ed. London: Butterworths.

Cochran KW, Doull J, Mazur M, DuBois KP. 1950. Acute toxicity of zirconium, columbium, strontium, lanthanum, cesium, tantalum and yttrium. Arch Ind Hyg Occup Med 1:637-650.

Minerals Yearbook 1975. 1977. Volume I. Metals, minerals, and fuels. Bureau of Mines. U.S. Department of the Interior. Washington DC: U.S. Government Printing Office.

OSHA. 1981. Occupational Safety and Health Admin. Occupational safety and health standards. Subpart 2--Toxic and hazardous substances. Code of federal regulations 29 (Part 1910.1000) pp. 673679.

Stokinger HE. 1981. Chapter 29. The metals. In: Patty's industrial hygiene and toxicology, 3rd ed., volume IIA, Toxicology. Clayton GD, Clayton FE, eds. New York: Wiley-Interscience, John Wiley & Sons, pp. 1493-2060.

TELLURIUM

MAMMALIAN TOXICITY SUMMARY

I.   INTRODUCTION

A.   Occurrence and Production

Tellurium occurs naturally in the free state but is usually associated with ores such as sylvanite [(AgAu)Te$_2$], black tellurium (AuPb$_2$),
hessite (Ag$_2$Te), and tetradymite (Bi$_2$Te$_3$) as tellurides. Tellurium is produced mainly from the residues of bismuth ores (Stokinger 1981; Browning
1969).

B.   Uses

Tellurium is used to improve metal alloys, as an additive in rubber, as a catalyst in the chemical industry, in electronic components, and
in "daylight" lamps (Stokinger 1981).

C.   Chemistry

Tellurium, atomic number 52, melts at 450°C and boils at 990°C
(Merck Index 1983). It is insoluble in water, hydrochloric acid, and carbon
disulfide; it is soluble in oxidizing acids and alkalies. Its chemistry is
similar to that of selenium. The oxide is almost insoluble in water and body
fluids (Stokinger 1981; Browning 1969). Te(OH)$_6$, telluric acid, is water
soluble with a strong tendency to polymerize; solutions become colloidal.
TeCl$_4$ is decomposed in water to TeO$_2$ and HCl (Merck Index 1983).

II. EXPOSURE AND EXPOSURE LIMITS

A.   Oral

Fatty food and some manufactured foods (processed food, baking
powder, instant beverages) are highest in tellurium. A volatile telluride
occurs in garlic. One report gave 0.73 mg/day as the tellurium content of a
hospital diet. Tellurium intake from drinking water is insignificant (Snyder
et al. 1975).

B.   Inhalation

There is no significant tellurium intake from ambient air. The
threshold limit value as a time-weighted average in workroom air for an 8-h
day is 0.1 mg Te/m^3 for tellurium (ACGIH 1983). The OSHA permissible
exposure limit is 0.1 mg/m^3 (OSHA 1981).

C.   Dermal

No information.

D. Total body Burden and Balance Information

Tellurium balance of 70-kg reference man is given by Snyder et al.
(1975) as follows:

Intake, mg/day		Losses, mg/day	
Food and fluids	0.6	Urine	0.53
		Feces	0.1
		Breath	0.01

III. TOXICOKINETICS

A. Absorption

Tellurium is poorly absorbed from the intestional tract; only ~ 25%
of the amount ingested is absorbed. It is not certain whether salts of
tellurium are absorbed through the skin, but tellurous acid is absorbed.
Absorption of tellurium produces a peculiar, unpleasant garlic-like odor of
the breath attributed to Me_2Te (Browning 1969; Stokinger 1981).

B. Distribution

1. General

Tellurium concentration is greatest in the kidneys, 3 to 5 times
that of the other storage organs, the heart, lungs, and spleen (Browning
1969). However, Snyder et al. (1975) list yellow marrow and adipose as
having the highest contents of tellurium, 11 and 2.2 mg, respectively.

2. Blood

The concentration of tellurium is greater in the plasma than in the
red blood cells and is about 0.5 to 7% the concentration in the urine
(Browning 1969). [Stokinger (1981) gives 0.2 to 1.0 μg/mL as the normal
urine concentration.]

C. Excretion

Tellurium excretion is greater by the urine than the feces when
given intravenously, but the reverse occurs when tellurium is given orally.
After 6 days, 20% of the intravenous dose and 3.7% of the oral dose had been
excreted. Some biliary excretion occurs (Browning 1969). To prevent the
breath odor, urine concentrations should be < 1 μg Te/L (Lauwreys 1983).

IV. EFFECTS

A. Acute and Other Short-Term Exposures

In humans, an accidental dose of sodium tellurite estimated to be
2 g was fatal in 2 of 3 cases. Cyanosis and garlic breath was a marked

feature of all 3 cases with the 2 who died also experiencing nausea and vomiting, followed by stupor, loss of consciousness, and death within 6 h (Keall et al. 1946; cited by Browning 1969).

In the internal organs, congestion, edema, and fatty degeneration especially in the liver were found. Tellurium given therapeutically in the form of a glucose suspension by intramuscular injection to treat syphilis caused no symptoms other than garlic breath and a metallic taste in the mouth (Fournier and Levadite 1926; cited by Browning 1969). In an early self experiment, ingestion of 0.4 to 0.88 g of K tellurite for 7 days resulted in drowsiness, anorexia, nausea, cardiac oppression, and the characteristic odor on the breath that lasted 7 wk (Hansen 1853; cited by Browning 1969). Occupational exposure to tellurium has resulted in odor on the breath and drowsiness at levels of 0.1 to 1.0 mg Te/m^3. In other industrial exposures, additional symptoms reported were nausea, anorexia, suppression of sweat, dry itching skin, and constipation or diarrhea (Browning 1969).

In animals, the lethal dose by injection for the dog was 0.5 g K tellurite; orally, 0.7 g caused drowsiness and vomiting. Guinea pigs died 48 h after intramuscular injection of 75 mg telluric oxide (Browning 1969).

B.   Chronic Exposure

Effects detected in animals after chronic exposures to tellurium or its compounds were garlic breath, digestive disturbances, lack of growth, emaciation, somnolence, and loss of hair. Elemental tellurium, especially when given orally, appears to be less toxic than its compounds; it reportedly had no effect at 1500 ppm in feed whereas tellurite and telluride caused toxic effects beginning at 25 to 50 ppm. Ducks exposed to TeCl$_4$ in their diet at 50 to 1000 ppm had reduced growth and heavy mortality at 500 and 1000 ppm by the second week. At 50 and 250 ppm, growth was not affected, but 50 to 70% mortality was reached by the fourth week (Carlton and Kelly 1967; cited by Browning 1969). Stokinger (1981) cites studies in which repeated oral dosing of animals with tellurium [compounds?] produced kidney and nerve damage.

C.   Biochemistry

1.   Effects on Enzymes

No information.

2.   Metabolism

The odor associated with tellurium is believed to be due to the formation of methyl telluride (Me$_2$Te) by synthesis within the body (Browning 1969).

3.   Antagonisms and Synergisms

Dosing with ascorbic acid (vitamin C) reduces or abolishes the Me$_2$Te odor of the breath temporarily. BAL (British anti-Lewisite, dimercaprol) enhances the toxicity of tellurium, perhaps by forming a complex (Browning 1969).

D.   Specific Organs and Systems

1.   Brain and Nervous Tissue

In ducks exposed to 50 to 1000 ppm tellurium tetrachloride, neural lesions were found in the brain cerebrum and gray-black discoloration of the intestine and kidneys, presumed to be deposits of elemental tellurium.

2.   Gastrointestinal Tract

Dogs given lethal doses of K tellurite had inflammation and hemorrhages of the digestive tract and intense hyperemia of the internal organs.

3.   Kidneys

Guinea pigs given lethal does of tellurium oxide had hemorrhage and necrosis of the kidneys. See also IV.D.1.

4.   Liver

In humans given fatal doses, congestion, edema, and fatty degeneration of the internal organs, especially the liver, was noted (Browning 1969).

E.   Teratogenicity

Hydrocephalus in the offspring occurs following the administration of tellurium to pregnant rats, especially on days 9 and 10 of gestation (Stokinger 1981; Shepard 1980).

F.   Mutagenicity

No information.

G.   Carcinogenicity

No information.

V.   REFERENCES

ACGIH.   1983.   TLVs® threshold limit values for chemical substances and physical agents in the work environment with intended changes for 1983-84. Cincinnati, Ohio: American Conference of Governmental Industrial Hygienists.

Beliles RP. 1981. Chapter 31. Phosphorus, selenium, and tellurium. In: Patty's industrial hygiene and toxicology, 3rd ed., volume IIa, Toxicology. Clayton GD, Clayton FE, eds. New York: Wiley-Interscience, John Wiley & Sons, pp. 2121-2140.

Lauwerys RR. 1983. Chapter II. Biological monitoring of exposure to inorganic and organometallic substances. In: Industrial chemical exposure: Guidelines for biological monitoring. Davis CA: Biomedical Publications, pp. 9-50.

Merck Index, 10th ed. 1983. Rahway, New Jersey: Merck & Co., Inc.

OSHA. 1981. Occupational Safety and Health Admin. Occupational safety and health standards. Subpart 2--Toxic and hazardous substances. Code of federal regulations 29 (Part 1910.1000) pp. 673-679.

Shepard TH. 1980. Catalog of teratogenic agents, 3rd ed. Baltimore: The Johns Hopkins University Press.

Snyder WS, Cook MJ, Nasset ES, Karhausen LR, Howells GP, Tipton IH. 1975. International Commission on Radiological Protection. Report of the task group on reference man. New York. ICRP Publication 23.

THALLIUM

MAMMALIAN TOXICITY SUMMARY

I. INTRODUCTION

A. Occurrence and Production

In the natural environment, thallium(I) is widely dispersed as an isomorphous replacement for K^+ in potassium feldspars and micas (silicates). The crustal abundance of thallium ($\sim$ 1 ppm) is largely due to these minerals in clays, soils, and granites. Thallium sulfide is found in widely varying amounts in sphalerite, galena, pyrite, and other sulfide mineral deposits. Some manganese oxide deposits are especially enriched (in the tenth of a percent range), and deep-sea manganese nodules usually contain hundreds of parts per million (ppm) of thallium. Most authorities believe thallium is transported in environmental waters as Tl^+. Under strong reducing conditions, thallium becomes enriched in the sediments. Under strongly oxidizing conditions, thallium(I) may be oxidized and precipitate and/or be adsorbed as Tl_2O_3 along with hydrated manganese and/or iron oxides. Thallium(I) is strongly fixed by soils.

The commercial recovery of thallium has always been from materials associated with sulfide ore processing; that is, dusts collected from roasting and smelting zinc, lead, copper, and iron sulfide concentrates, and from purification of zinc sulfate solutions in electrolytic zinc refining and lithopone production. The only identified U.S. producer of commercial thallium and thallium sulfate, at least since the 1930's, has been ASARCO Inc., at its cadmium refinery in Denver. Most of the thallium entering U.S. zinc and cadmium refineries is not recovered as such but leaves in product zinc or is recycled, ultimately ending up in some undefined waste stream (Smith and Carson 1977).

B. Uses

Thallium has few uses. Consumption is about 0.5 ton/year and is not well defined. The major uses are apparently in electrical and electronic applications. The former major use (perhaps 5 tons/year at its peak) of thallium compounds in rodenticides (chiefly western ground squirrel control) and insecticides was terminated in the United States in 1972 because of accidental or secondary poisoning of wild animals and birds (especially carnivores). Children were often severely poisoned by ingesting the attractive baits. The use of thallium acetate as a cosmetic depilatory (a 7% cream) around 1930, as well as its use for about 50 years as a therapeutic epilant (one-time dose of 8 mg/kg in children) in the treatment of fungal scalp infections, was often accompanied by severe poisonings and fatalities.

C. Chemical and Physical Properties

Thallium, atomic number 81, forms compounds in both the monovalent and trivalent state. In ionic radius and chemical behavior, Tl^+ resembles the alkali metal cations. TlOH is water soluble. However, thallium(I) forms sparingly soluble compounds: sulfides, iodides, chlorides, chromates, etc., analogous to the heavy metals of Group Ib, Cu^+, Ag^+ and Au^+, and to its nearest neighbors in the sixth period, Hg_2^{2+} and Pb^{2+}. The standard potential of the Tl^{3+}/Tl^+ couple apparently favors Tl^+. The Tl^{3+} is assumed attainable only through oxidation by very powerful oxidizing agents in very acid media. Recent arguments, however, propose that formation of Tl(III) chloride complexes in high chloride media would favor the trivalent state. Tl(III) inorganic compounds are generally more water soluble, but hydrolysis to colloidal Tl_2O_3 and $TlOH^{2+}$ is extensive. Tl(III) forms more stable organic compounds and has been methylated _in vitro_ by methyl vitamin B_{12}.

II. EXPOSURE AND EXPOSURE LIMITS

A. Oral

Usual levels for thallium in freshwater are 0.01 to 14 µg/L (Smith and Carson 1977), but they seldom exceed 30 µg/L and tapwater seldom exceeds 0.3 µg/L (EPA 1980). USEPA (1980) recommended a criterion of 13 µg Tl/L for drinking water based on an acceptable daily intake of 37.1 µg Tl. Dietary intake and excretion is on the order of 2 µg/day. Levels > 2 ppm in the ash of terrestrial plants are abnormal, occurring only in high mineralized areas. Freshwater animals appear to bioconcentrate thallium with bioconcentration factors < 20. However, a concentration factor of > 700 has been reported for a marine organism.

Biota in thallium-contaminated areas currently have thallium levels (< 3 ppm) that could be high enough to cause toxic symptoms if they constituted the entire diet of a mammal.

Accidental poisonings have occurred in the past from use of thallium rodenticides, but their use has been banned in the United States (Smith and Carson 1977).

B. Inhalation

In their extensive review of environmental exposure to thallium, Smith and Carson (1977) found ambient air concentrations reported only for Chadron, Nebraska. The values were 0.04 to 0.48 ng Tl/m^3.

The U.S. permissible exposure limit (OSHA Standard) (OSHA 1981) and threshold limit value (ACGIH 1983) for thallium compounds in workplace air is 0.1 mg/m^3 (with a skin exposure warning), largely based on analogy with other highly toxic compounds. The USSR stipulates 0.01 mg/m^3 as a ceiling limit (Smith and Carson 1977).

C. Dermal

Dermal exposures to thallium may occur while handling solutions used in mineralogical analysis and other thallium preparations. Tl^+, unlike Na^+ and K^+, can penetrate intact skin, producing serious poisoning (Stokinger 1981).

D. Body Burden and Balance Information

Based on extensive tissue data for a 16-year-old German male, Smith and Carson (1977) calculated a human body burden of about 2 ppb (about 0.14 mg/70 kg man).

Smith and Carson's (1977) estimate of 2 µg/day in diet and excreta is close to that of Snyder et al. (1975) for Reference Man:

Intake, µg		Losses, µg	
Food and fluids	1.5	Urine	0.5*
Airborne	0.05	Feces	1.0
		Hair	0.0001

---

* Urine samples of 10 normal subjects, analyzed by an extraction-spectrometric method, had 0.02 to 1.0 µg Tl/kg. Another source estimated that 0.6 to 2.5 µg Tl is excreted in the urine daily by normal subjects (1.5 kg urine/day) (Smith ' and Carson 1977).

III. TOXICOKINETICS

A. Absorption

Thallium absorption is apparently complete by any route of adminis-tration. Like potassium, thallium is distributed in the intracellular space of most tissues. Transplacental absorption of maternally toxic doses can kill fetuses (Smith and Carson 1977).

B. Distribution

1. General

After acute poisoning, the kidneys, especially the renal medulla, usually contain the highest thallium concentrations. One suicidal poisoning (death on the 5th day) showed the highest thallium concentration in the liver. An experiment with goats showed that ^{201}Tl was most concentrated in the heart for the first 10 min after injection; but for 25 min to 243 days, it was highest in the kidneys. No information was found by Smith and Carson (1977) on thallium in the tissues of normals. In the final stages of fatal poisoning, thallium appears in all organs and the tissue concentrations tend to equalize.

2. Blood

About 70% of the thallium in the blood is in the red blood cells (Lauwerys 1983).

Blood is not a reliable indicator of thallium exposure. Only a small fraction of the body burden of thallium in poisoned subjects is in the plasma and extracellular fluid. In about 2 h after entry, practically all the thallium has been transported to the cells. In one adult, 2 days after poisoning, a concentration of 5.5 mg Tl/kg blood was reported (Smith and Carson 1977).

C. Excretion

Most studies indicate that macro amounts of absorbed thallium, as opposed to radiotracer doses, are excreted predominantly in the urine rather than in the feces (Smith and Carson 1977). Goetz et al. (1981) suggested that saliva might be useful as an indicator of thallium exposure after they detected high concentrations of radiothallium in the saliva of rats after an acute exposure. They found that biliary excretion by the liver is not important.

Determination of thallium in urine is probably a better indicator of exposure than blood. Urine levels of ≥ 50 µg Tl/L indicate poisoning (Kemper 1979; cited by Schaller et al. 1980).

Schaller et al. (1980) used flameless atomic absorption spectrometry (acceptable quality control and methodology) to determine thallium in urine of normals and of workers exposed to thallium in a raw material used for the iron clinker at three West German cement plants. Normal values were 1.1 µg Tl/g creatinine excreted. None of the cement workers exhibited signs of thallium intoxication although only 28 to 67% of the urine values (< 0.3 to 6.3 µg Tl/g creatinine) fell within the normal range.

Brockhaus et al. (1981a) considered 0.8 µg Tl/L as the upper normal limit for thallium in urine. They monitored urine concentrations of a population living around a cement plant emitting thallium-containing dust for several years. Subjects having possibly thallium-related health disorders (not given) in 1979, when the thallium problem was recognized, had the following distribution in 1980 urine concentrations:

µg/L Urine	Cumulative % of Population
> 40	1.4
> 20	10.8
> 5	44.6
> 0.8	86.5

Brockhaus et al. (1981b) (good analytical methodology) correlated the incidence of typical thallium poisoning symptoms with urine concentrations in 1,265 persons whose gardens had been contaminated by thallium-bearing dusts. There was a significant correlation between exposure as evidenced by thallium urine concentrations and the following symptoms: sleep disorders (9.8% in > 20 µg/L group compared to 2.1% in < 2 µg/L group) and neurological symptoms such as paresthesia and pain in the muscles and joints (41.2% for the high exposure versus 14.5% for the low-exposure group).

Thallium in hair as a measure of exposure had a similar exposure-response relationship. However, thallium in urine is probably a more reliable parameter for monitoring current exposure.

IV. EFFECTS

  A. Acute and Other Short-Term Exposures

       Major reviews of human thallium poisoning were published by Munch (1937), Heyroth (1947), Prick et al. (1955), and Cavanagh et al. (1974). These reviews were used by Smith and Carson (1977) in their appraisal of environmental exposure to thallium. More recently, Stokinger (1981) [based on literature up to 1979] has briefly reviewed thallium toxicity.

       For humans, doses of 14 mg/kg and above are fatal. The minimum lethal dose in birds is about 15 to 50 mg/kg. In mammals, the LD_{50}'s of thallium(I) inorganic compounds range from 15.8 to 71 mg/kg; of thallium(III) inorganic compounds, 5.66 to 72 mg/kg. The least toxic doses were for the oxides by the intraperitoneal route. The LD_{LO}'s for thallium(I) compounds range from 7 to 45 mg/kg; for thallium(III) compounds, 5 to 62 mg/kg. Lethal doses of dimethylthallium(III) bromide to mice are 5 to 6 mg/kg (Smith and Carson 1977).

       In mammals, toxic effects are usually delayed for 12 h to 2 days even when a fatal dose is given. Neurological symptoms do not appear for 2 to 5 days. The systemic effects in humans of acute thallium poisoning include: gastrointestinal symptoms, limb pain and paralysis, polyneuritis, high blood pressure, optic nerve atrophy and blindness, psychic excitement, liver and kidney damage, lymphocytosis and other blood changes, and (at least 10 days after poisoning) hair fall.* Without known association of the patient with possible sources, the systemic effects are difficult to ascribe to thallium poisoning until alopecia occurs. Even with thorough investigation, numerous diagnoses have been listed on the death certificates of thallium-poisoned murder victims; but to recognize symptoms as being due to an environmentally caused thallium intoxication would be even more difficult. Symptoms in animals are similar. Weight loss is frequently noted. When occupational poisonings have occurred, despite protective clothing, the effects have usually been severe (Smith and Carson 1977).

       Severe hemorrhagic gastroenteritis occurs within 24 h of acute oral toxic doses of thallium. Delirium and convulsions, then severe depression and coma may follow. The usual cause of death is respiratory depression with pneumonia or respiratory paralysis. Other deaths from thallium poisoning have been attributed to cardiac disturbance, dehydration, and progressive impairment of the brain and vagus nerve (Smith and Carson 1977).

_____

* All of the symptoms from the older reviews mentioned above are summarized on pages 202 to 203 of Smith and Carson (1977). The horizontal white bands or white cross lines (Mees' lines) on the nails may be a late sign but are not characteristic of thallium poisoning.

Severe parenchymatous changes were found in three boys fatally poisoned by thallium, especially in the boy who had survived for 3 days. Fatty
degeneration of the heart was especially prominent in the latter. All three
showed "tabby cat" striation of the left ventricle (both ventricles in one
boy). The liver lobules showed fatty degeneration and nephrosis. In a
series of California poisonings, seven autopsies revealed stomatitis, fatty
liver and central liver necrosis, lung edema, meningeal congestion, renal
damage, gastroenteritis, and widespread degeneration of the nerve cells and
axons in the brain. The only endocrine gland affected was the adrenal with
marked hyperemia, small medullary hemorrhages, and areas of necrosis and
nuclear degeneration (Smith and Carson 1977).

B. Chronic Exposure

According to Pritschow (1959), chronic thallium poisoning from multiple doses over a long interval, shows a long latent period, and produces
either severe or mild symptoms.

Few of the reported human and mammalian studies of thallium toxicity lend themselves to any conclusions about the dangers of very low
chronic intakes. One equivocal human study implies that minor toxic effects
(alopecia) arise at daily intakes of about 10 to 20 µg/day. In that study,
12 of 190 Canadian refinery workers, who experienced alopecia areata, were
not known to be exposed to thallium, yet their urine concentrations were 8 to
15 µg Tl/L and similar concentrations were found in unaffected workers. In
four cases of definite occupational thallium poisoning, which initially
exhibited sensory changes in the extremities, minor and uncertain signs of
polyneuritis, and alopecia, the highest urine concentration was 380 µg Tl/L
(Smith and Carson 1977). However, urinary thallium concentrations in ∿ 60
workers in a plant manufacturing Tl-Mg alloy ranged from < 50 µg/L (a
provisional "alerting level") to 238 µg/L. After improved industrial
hygiene, all urine thallium concentrations (except one) were < 4 µg/L after 3
years. No illness was ever attributed to thallium (Stokinger 1981).

In humans, alopecia is the hallmark of long-term thallium poisoning
with hair loss beginning within about 10 days and epilation being complete
within a month (Stokinger 1981). However, alopecia does not always occur
even after severe poisoning. Polyneuritis, alopecia with eyebrow
involvement, and optic nerve atrophy are diagnostic of thallium poisoning,
but alopecia and polyneuritis usually do not occur for at least 10 days.
Sleep disturbances may be characteristic with patients dozing by day but
unable to sleep at night, even when sedated with opiates (Smith and
Carson 1977).

Rats fed 5 and 15 ppm thallous acetate (78% Tl) in the diet for 15
weeks grew as well as the controls, but alopecia occurred at 15 ppm (equiva-
lent to ∿ 0.8 to 2 mg Tl/kg/day, total dose ∿ 24 mg Tl).* Concentrations of

---

*   This is the experiment upon which the 1980 EPA drinking water criterion
    is based (USEPA 1980).

$\geq$ 50 ppm thallous acetate in the diet had killed 60 to 100% of the male weanling rats within 10 days; all eventually died. Fed thallic oxide (89% Tl) at 20 ppm in the diet for 15 weeks (total dose $\sim$ 36 mg Tl/rat), females grew as well as controls, and males showed significant body weight depression but all survived (two females died). Hair follicle atrophy and alopecia were seen in both sexes (Smith and Carson 1977).

Rabbits receiving daily subcutaneous or oral doses of thallium carbonate (total dose $\sim$ 80 mg Tl or 0.4 mg/day) or thallium sulfate (total dose $\sim$ 100 mg Tl or $\sim$ 0.6 mg/day) for 6 mo began to show behavioral (aggressiveness) and neurological symptoms (retardation, hind limb paralysis) at 5 mo. Hyperglobulinemia and hypoalbuminemia were also observed (Smith and Carson 1977).

C. Biochemistry

1. Effects on Enzymes

At low concentrations, Tl^+ has an affinity for certain enzymes and an activating ability 10 times that of K^+; its effects on Na^+, K^+ - activated ATPase of various tissues are those that have been most thoroughly studied. Thallium appears to uncouple mitochondrial oxidative phosphorylation (the reaction of adenosine diphosphate with inorganic phosphate to produce ATP). Toxic doses adversely affect protein synthesis and cause disaggregation of ribosomes. Thallium salts inhibit several enzymes such as succinic acid dehydrogenase and alkaline phosphatase, which plays a major role in bone formation (Smith and Carson 1977).

2. Metabolism

No reports of chemical changes of thallous ions in the body were found.

3. Antagonisms and Synergisms

Selenium- and sulfur-containing compounds offer some protection against thallium toxicity, but the only antidote that has given uniformly favorable results in animal experiments and clinical trials is Prussian blue [potassium ferrihexacyanoferrate(II)]. Thallium acetate aggravates the symptoms of dihydrotachysterol overdosage (including nephrocalcinosis). The salt also induces renal calcification when given with sodium acetate, sodium citrate, sodium dihydrogen phosphate, and disodium hydrogen phosphate. Thallium given together with zinc or barium decalcifies osseous tissue. Thallium induces riboflavin deficiency in rats, leading to optic nerve damage (Smith and Carson 1977).

4. Physiological Requirements

None.

D. Specific Organs and Systems

   1. Eye and Nervous System

        The most frequent sign of the ocular effects of thallium poisoning
is retrobulbar neuritis with reduction of vision and central scotoma. Rarely
edema of the papilla nervi optici occurs. Vision may recover partially but
slowly.

        Doses sufficient to cause hair loss in animals produced cataracts.
Lid inflammation, intraocular hemorrhage, retrobulbar neuritis, and optic
nerve atrophy were also reported. Nerve cells of humans and animals exhibit
axonal degeneration and demyelination after high thallium doses (Smith and
Carson 1977).

        Cavanagh (1973) and Cavanagh et al. (1974) reviewed the peripheral
neuropathy caused by thallium poisoning. Long nerves (in the leg and extra-
ocular muscles) are affected earlier than short nerves. Larger diameter fibers of sensory nerves are attacked selectively. The only central nervous
system changes are chromatolysis of motor nerve cells and degeneration of the
gracile tracts in the dorsal spinal cord. Apparently, all cranial nerves except I and VIII are affected by thallium poisoning. Cranial nerve involvement usually occurs before optic neuropathy.

   2. Skin and Hair

        Skin disorders caused by thallium poisoning include scaly erythema,
eruptions, keratinization of the epithelium, ecchymoses, and petechia (Smith
and Carson 1977).

        KRS-5 crystals (42 mole % TlBr + 58 mole % TlI, water solubility
0.2 mg/mL at 20°C) used for infrared transmission, may come into contact with
human skin during ordinary handling of the large single crystals in the spectroscopy laboratory or when reflection techniques are used to study living
skin. Oertel and Newmann (1972; cited by Smith and Carson 1977) found no
skin hypersensitivity among 20 guinea pigs (10 controls) to powdered KRS-5
exposed 6 h once a week for 3 weeks to a thin layer of KRS-5.

        Some investigators believe the hair loss is due to action of
thallium ions on the central or sympathetic nervous system. The theory that
systemic hair loss from thallium poisoning is due to autonomic abnormality is
favored by the absence of local skin changes and the impermanence of the
alopecia. On the other hand, "sensory hairs" in animals are spared; and when
autotransplants of denervated skin are exposed to thallium, the hair is lost.
Other investigators have described atrophy and/or hyperplasia in structures
of the skin and hair follicles and believe that the direct action of thallium
on these structures causes alopecia (Smith and Carson 1977).

   3. Liver

        Fatty infiltration (a dystrophic change) is often found in livers
of humans and animals severely or fatally poisoned by thallium. In the liver

(as well as the kidney and intestine), mitochondria most frequently showed degenerative changes: swelling (thought to be due to thallium accumulation within them), partial or total loss of cristae and mitochondrial granules, increased size and density of the latter occasionally with electronlucent cores, and increased numbers and stacking of mitochondrial cristae (Smith and Carson 1977).

4. Kidneys

Pritschow (1959) reviewed the histological literature according to the changes--parenchymatous, degeneration, fatty infiltration, hemorrhage, congestion, edema, inflammation, karyolysis, vacuolation, necrosis, and atrophy--in the various organs of thallium-poisoned humans and animals. The digestive system, nervous system, and kidneys contain the predominant lesions.

Rats poisoned with subcutaneous injections of up to 50 mg thallium acetate acutely, subacutely, or chronically consistently showed changes in the kidney, liver, and intestine under the electron microscope. In the kidney, swollen tubule cells, dilated or partly ruptured endoplasmic reticulum; intraluminal casts in the tubules; and increased numbers of autophagic vacuoles, lysosomes, lipid droplets, and residual bodies in the tubular cells were observed. Autophagic vacuoles, lysosomes, lipid droplets, and mitochondrial changes are not specific for thallium poisoning and are similar to cellular injury sequelae of many agents (Smith and Carson 1977).

5. Gastrointestinal Tract

Histopathological changes often occur in the intestine. Rats acutely poisoned with up to 50 mg thallium acetate per kilogram subcutaneously showed enteritis and severe colitis of the intestine. Fatally poisoned humans showed congestion of the internal organs, stomatitis, and punctate hemorrhages along the digestive tract (Smith and Carson 1977).

6. Heart and Blood Pressure

Heart injury by thallium poisoning has been evinced by sinus tachycardia, bradycardia, angina pectoris, and reduced or increased blood pressure. Injury to the vagus by thallium produces electrocardiographic changes.

Acute thallium poisoning has a hypotensive effect. The blood pressure is hypertensive, however, at longer terms. Thallium, like potassium, depresses the heart directly and by action on the parasympathetic nervous system. Hypotension induced by thallium masks a stimulating action on the sympathetic nervous system. The hypotension disappears on vagotomy; hypertension, occurring simultaneously with renal vasoconstriction, appears by blocking the parasympathetics with atropine (Smith and Carson 1977).

7. Blood

Chronically poisoned rats show a transitory hyperglycemia before blood sugar lowering. Hyperglycemia follows intravenous injection of a

thallium salt. Subcutaneous injections given repeatedly increase blood levels of potassium and cholesterol and decrease calcium. Erythropenia and
leukocytosis are only occasionally observed. Characteristic changes in the
formed elements of the blood were not observed in the acute cases in
California, but Flamm (1926; cited by Smith and Carson, 1977) noted a 21.7%
increase in leukocytes, a 39% increase in lymphocytes, and an 80% increase in
eosinophiles in the blood of children 16 days after receiving an epilating
dose of thallium.

8. Bones

Growth retardation by thallium in warm-blooded animals may be due
to failure of the body to store calcium. The effect resembles rickets. The
bones of young rats poisoned by thallium are soft, curved, poorly calcified,
and show fusiform osteoid tissue deposits. The marrow is fibrous. Sixty-
eight percent of rats given 0.2 mg thallous acetate daily developed bony lesions within 4 to 6 weeks. Thallium decreased the ash content of rat teeth,
but did not change the phosphorus content markedly. Small doses of thallium
increased calcium content; large doses decreased it. Phosphorus retention
may be impaired during the first stage of poisoning, but calcium stores decrease gradually with advancing intoxication (Smith and Carson 1977).

E. Teratogenicity

The deleterious effects of thallium compounds on reproductive organs have been judged to be no worse than those of any other general cellular
poison; but fetal development is altered and fetal mortality is increased.
Chicks exhibit achondroplasia, extreme leg-bone curvature and shortening,
parrot beak, and other abnormalities. Rats show non-ossification of the
phalanges and vertebral bodies when the mother is given thallium while fed a
low potassium diet (thallium sulfate at 2.5 mg/kg/day on days 12, 13, and
14). Mucopolysaccharide synthesis is inhibited so that columnar cartilage of
the long-bone is hypoplastic and the calcifying zone is defective. Sea
urchin egg development is inhibited by 20 to 200 ppm thallous or thallic
chloride, but 48 and 240 ppm do not affect tadpole development in the egg
(Smith and Carson 1977).

F. Mutagenicity

Thallium salts have marked antimitotic activity on mammalian,
avian, and plant cells. Thallium(I) is the most active heavy metal ion found
for breaking chromosomes of pea plants. Thallous acetate solutions retarded
both mitosis and meiosis in mosquitoes when the larvae and nymphs were
treated (Smith and Carson 1977).

G. Carcinogenicity

Six papers published in the 1920's reported that rats given chronic
oral doses of thallium salts showed inflammatory proliferative lesions in
parts of the gastrointestinal tract. Champy et al. (1958; cited by Smith and
Carson 1977) reported that chronic oral or cutaneous dosing of mice with
thallium salts caused degeneration, papillomas, precancerous lesions, and

cancers of the female genital tract. Experimental details and the incidence of lesions and mortality were not given (Smith and Carson 1977).

V. REFERENCES

ACGIH. 1983. TLVs® Threshold limit values for chemical substances and physical agents in the work environment with intended changes for 1983-84. Cincinnati, Ohio: American Conference of Governmental Industrial Hygienists.

Brockhaus A, Dolgner R, Ewers U, Freier I, Jermann E, Wiegand H. 1981a. Repeated surveillance of exposure to thallium in a population living around a cement plant emitting thallium containing dust. Int Conf, 3rd. Heavy Met Environ 482-485.

Brockhaus A, Dolgner R, Ewers U, Kraemer U, Soddemann H, Wiegand H. 1981b. Intake and health effects of thallium among a population living in the vicinity of a cement plant emitting thallium containing dust. Int Arch Occup Environ Health 48(4):375-389.

Cavanaugh JB. 1973. Peripheral neuropathy caused by chemical agents. Crit Rev Toxicol 2(3):365-417.

Cavanagh JB, Fuller NH, Johnson HRM, Rudge P. 1974. The effects of thallium salts, with particular reference to the nervous system changes. Quart J Med 43:293-319.

Goetz L, Sabbioni E, Marafante E, Edel-Rade J, Birattari C, Bonardi M. 1981. Biochemical studies of current environmental levels of trace elements: cyclotron production of radiothallium and its use for metabolic investigations on laboratory animals. J Radioanal Chem 67(1):183-192.

Heyroth FF. 1947. Thallium: a review and summary of medical literature. Public Health Rep (US), Suppl 197:1-23.

Lauwerys RR. 1983. Chapter II. Biological monitoring of exposure to inorganic and organometallic substances. In: Industrial chemical exposure: Guidelines for biological monitoring. Davis, CA: Biomedical Publications, pp. 9-50.

Munch JC, Silver J. 1931. The pharmacology of thallium and its use in rodent control. USDA Tech Bull 238:1-28.

OSHA. 1981. Occupational Health and Safety Administration. Occupational safety and health standards. Subpart 2--Toxic and hazardous substances. Code of federal regulations 29 (Part 1910.1000): 673-679.

Prick, JJG, Smitt WGS, Muller L. 1955. Thallium poisoning. Elsevier: Amsterdam.

Pritschow AL. 1959. A study of the distribution of thallium in tissues, blood, urine and feces. Ph.D. dissertation, University of Oklahoma, Oklahoma City, Oklahoma.

Schaller KH, Manke G, Raithel HJ, Buehlmeyer G, Schmidt M, Valentin H. 1980. Investigations of thallium-exposed workers in cement factories. Int Arch Occup Environ Health 47(3):223-231.

Smith IC, Carson BL. 1977. Trace metals in the environment: volume 1 - thallium. Ann Arbor, Michigan: Ann Arbor Science Publishers, Inc., pp. 394.

Snyder WS, Cook MJ, Nasset ES, Karhausen LR, Howells GP, Tipton IH. 1975. International Commission on Radiological Protection. Report of the task group on reference man. New York. ICRP Publication 23.

Stokinger HE. 1981. Chapter 29. The metals. In: Patty's industrial hygiene and toxicology. 3rd ed. Volume 2A. Clayton GD, Clayton FE, eds. New York: A Wiley-Interscience Publication. John Wiley and Sons, pp. 1493-2060.

USEPA. 1980. Ambient water qualty criteria for thallium. Springfield, Virginia: National Technical Information Service. PB81-117848.

THORIUM

MAMMALIAM TOXICITY SUMMARY

I. INTRODUCTION

Stokinger (1981) largely based his review of thorium in Patty's Industrial Hygiene and Toxicology on Albert's 1966 book, Thorium: Its Industrial Hygiene Aspects.

The naturally radioactive element thorium comprises 0.001 to 0.002 wt % of the earth's crust. Thorium is a by-product of uranium recovery from uraininite [essentially UO_2] and is recovered chiefly from monazite [$(Ce,La,Y,Th)PO_4$], the principal ore of rare earth elements (Katzin 1983). Thorianite [$(Th,U)O_2$] is another important ore source of thorium and uranium. In the United States, thorium is produced as a by-product of mining beach sands including monazite in Florida for titanium and rare earths (Katzin 1983).

Thorium is an important source of ^{233}U in the breeder reactor according to these reactions:

$$^{232}Th(n,\gamma) \; ^{233}Th; \; ^{233}Th \xrightarrow{-\beta} \; ^{233}Pa \xrightarrow{-\beta} \; ^{233}U$$

Thorium production capacity will exceed demand until the ^{233}U-^{232}Th breeding cycle for nuclear fuels becomes commercially important (Katzin 1983).

Commercial uses of thorium include the manufacture of incandescent gas mantles, metallurgy (especially alloys with magnesium and as a reducing agent), refractories, electronics, and catalysts in organic syntheses (Merck Index 1983; Stokinger 1981).

The chemistry of thorium, atomic number 90, resembles that of the rare earths. Hydrolysis of thorium(IV) salts is common, but high dilution is required before precipitation of the hydroxides occurs (Stokinger 1981).

Thorium is grouped with the actinides; but unlike cerium, the corresponding first element of the lanthanides, thorium has only a tetrapositive stable state. Its long half-life makes ^{232}Th a minimal radiation hazard except in tonnage amounts.

II. EXPOSURE AND EXPOSURE LIMITS

A. Oral

Major oral exposures to thorium have occurred from medical use of the radiopaque medium Thorotrast or Thorium X (Stokinger 1981). There are no criteria for thorium in water.

B. Inhalation

Exposures to significant amounts of thorium are chiefly through occupational exposure during the production and use of thorium compounds and alloys, during the casting and machining of alloy parts, and during welding with thorium electrodes. The maximum permissible concentration in air is 1 x 10^{-6} $\mu Ci/mL$ air (Sittig 1981).

C. Dermal

Exposures of the skin are most likely for workers handling thorium compounds.

D. Total Body Burden or Balance Information

According to Snyder et al. (1975), 70-kg Reference Man ingests 3 μg Th/day in food and fluids and excretes 0.1 and 2.9 $\mu g/day$ in urine and feces, respectively.

III. TOXICOKINETICS

Most thorium entering the body as an inorganic salt is hydrolyzed and polymerized, giving colloidal particles that are phagocytized by cells of the reticuloendothelial system, liver, spleen, and bone marrow (Stokinger 1981).

A. Absorption

1. Gastrointestinal

Only 0.001% and 0.05% of thorium ingested as the nitrate is absorbed at doses of 500 to 800 mg/kg and 5 mg/kg, respectively (Stokinger 1981).

2. Lung

Doses of thorium administered intratracheally remain in the lung. After inhalation of $\sim$ 1-μm particles of thoria (ThO_2) dusts at $\sim$ 50 times the maximum permissible concentration for 1 yr, 98% of the body burden was in the lungs and pulmonary lymph nodes (Stokinger 1981).

3. Dermal

No information found.

D. Distribution

Six to seven years after inhalation exposure to ThO_2 dust, 97.7% of the total body burden was in the pulmonary lymph nodes, 2.1% in the lungs, 0.003% in the femur, and < 0.003 in the kidney, spleen, and liver (Stokinger 1981). No information was found on thorium in blood or adipose tissue.

C. Excretion

Fecal excretion of polymeric thorium compounds via bile from the liver predominates. Thorium in radiotracer (carrier-free) doses, however, remains monomeric and is predominantly excreted in the urine (Stokinger 1981).

IV. EFFECTS

A. Acute and Other Short-Term Exposures

Thorium salts exhibit very low acute toxicity by all routes studied.

1. Oral

A dog exposed to $Th(NO_3)_4 \cdot 4H_2O$ (1 g NO_3/kg/day) for 46 days showed 15% growth depression but no other toxic effect. Levels of 3% nitrate in the diet were required to cause growth depression in rats (Stokinger 1981).

2. Inhalation

Abnormal leukocytes were the only sign of toxicity in dogs exposed for 2 to 10 wk to thorium fluoride, oxalate, oxide, or nitrate at 11, 26, 51, or 76 mg/m³, respectively. Monocytes had atypical nuclei; abnormal lymphocytes were like those seen in toxic states, and old polymorphonuclear leukocytes were observed (Stokinger 1981).

3. Dermal

No information.

B. Chronic

Radiotoxicity, not chemical toxicity, is the chief concern of long-term exposure. Tumors develop at various sites in patients exposed many years after use of Thorotrast as a radiopaque medium (see Part IV.G.). Diseases attributed to thorium exposure in the workplace were not found in 693 former employees or 84 employees in 1955 of a thorium refinery. Laboratory animals exposed to 5 mg ThO_2/m³ air for 1 yr showed no observable adverse effects (Stokinger 1981).

C. Biochemistry

1. Effects on Enzymes

Thorium inhibits amylase and blood phosphatase (Gould 1936).

2. Metabolism

Thorium strongly binds cortical bone glucoprotein and sialoprotein (mainly due to binding with aspartic or glutamic acid). The chondroitin sulfate protein of cortical bone can also bind thorium (Stokinger 1981).

3. Antagonisms and Synergisms

Stokinger (1981) does not elaborate on the statement that thorium interacts with zinc in the rat prostate.

D. Specific Organs and Systems

1. Blood

See Part IV.A.2. for a description of the abnormal leukocytes observed after thorium inhalation exposure.

2. Other

Cancers of the blood vessels, liver, kidney, and other organs have appeared 11 to 37 yr after exposure to thorium as a radiopraque medium (Stokinger 1981) (see Part IV.G.).

E. Teratogenicity

No information.

F. Mutagenicity

No information.

G. Carcinogenicity

Verhaak et al. (1974; cited by Stokinger 1981) reported that 45 cases of tumors of the renal pelvis were reported in western Europe from 1949 to 1974. The mean latency period after Thorotrast pyelography was 27 yr. The severity of the neoplastic effect was unrelated to dose or time. Tumors are seen in many other body sites but not the bone. Only hemangioepithelioma of the liver was observed.

V. REFERENCES

Albert RE. 1966. Thorium: Its industrial hygiene aspects, New York: Academic Press, 222 pp.

Gould BS. 1936. Effects of thorium, zirconium, titanium and cerium on enzyme action. Proc Soc Exp Biol Med 34:381-385.

Katzin LI. 1983. Thorium and thorium compounds. In: Kirk-Othmer encyclopedia of chemical technology, Volume 22. New York, New York: Wiley-Interscience, John Wiley & Sons, pp. 989-1002.

Merck Index, 10th ed. 1983. Rahway, New Jersey: Merck & Co., Inc.

Sittig M.   1981.   Handbook of toxic and hazardous chemicals.   Park Ridge, New Jersey: Noyes Publication.

Snyder WS, Cook MJ, Nasset ES, Karhausen LR, Howells GP, Tipton IH.   1975. International Commission on Radiological Protection.   Report of the task group on reference man.   New York.   ICRP Publication 23.

Stokinger HE.   1981.   Chapter 29.   The metals.   In: Patty's industrial hygiene and toxicology.   3rd ed.   Volume 2A.   Clayton GD, Clayton FE, eds. New York: A Wiley-Interscience Publication.   John Wiley and Sons, pp. 1403-2060.

TIN

MAMMALIAN TOXICITY SUMMARY

I. INTRODUCTION

A. Occurrence and Production

Tin is found in a number of minerals. Cassiterite (SnO_2), stannite (Cu_2FeSnS_4), and teallite ($PbZnSnS_2$) are of commercial importance. Tin is produced by smelting to the oxide (if necessary), then reducing with carbon or reducing gases. The crude metal is then purified with heat or electrolytically (Stokinger 1981). There is one smelter in the United States in Texas City, Texas, which operates on imported tin concentrates.

B. Uses

Most tin is used as the metal as a protective coating (tin plate). Most of the rest goes into alloys, including solder, bronze, brass, babbitt metal, and pewter. Tin compounds are used as dyes, pigments, ceramics, bleaching agents, flame retardants, and fluoride sources in toothpaste. Organotins are used as antimicrobials, in antifouling paints, and as heat stabilizers in poly(vinyl chloride) polymers (Stokinger 1981).

C. Chemical and Physical Properties

Stannous compounds (+2 valence) are more important than stannic (+4) ones. Many salts are quite soluble. The metal becomes coated with a thin layer of SnO_2, which blocks further reaction (Stokinger 1981).

II. EXPOSURE AND EXPOSURE LIMITS

A. Oral

One study reported tin levels in water supplies in 42 U.S. cities as 1.1 to 2.2 µg/L (Beeson et al. 1976; cited by NAS 1977). Daily intake from food is much greater, with estimates ranging from 0.2 to 10 mg/day (Piscator 1977). SnF_2 in toothpaste is another oral source. Tin in food may be augmented by tin leached from unlacquered cans (Snyder et al. 1975).

B. Inhalation

The general population may inhale up to 7 µg Sn/day from ambient air (Snyder et al. 1975). The threshold limit value, as a time-weighted average in workroom air for an 8-h day is 2 mg Sn/m^3 for tin metal and inorganic compounds; the short-term exposure limit is 4 mg/m^3 for 15 min. For organotin compounds, the values are 0.1 and 0.2 mg Sn/m^3, respectively (ACGIH 1983). The OSHA permissible exposure limit is 2 mg/m^3 for inorganic tin compounds except oxides (OSHA 1981), but NIOSH (1976) has recommended a level of 0.1 mg Sn/m^3 for organotin compounds.

C. Dermal

Organotins are hazardous to skin (ACGIH 1983), but there is little likelihood for significant contact by the general public.

D. Total Body Burden and Balance Information

According to Snyder et al. (1975), the tin balance for the 70-kg reference man in mg/day is: Intake from food and fluids 4 and airborne 0.34×10^{-3}; losses from urine 0.02, in feces 3.5, and in sweat 0.5.

III. TOXICOKINETICS

A. Absorption

There are no data on inhalation. Ingested inorganic tin is poorly absorbed, on the order of 1%. Organotin compounds are better absorbed from the gut. Also, short-chain alkyltin compounds are absorbed through the skin (Piscator 1977).

B. Distribution

1. General

Inorganic tin is initially concentrated in the kidneys and liver, but soon shifts to the bones. Organotins are concentrated in the liver, and, for some compounds, kidneys and brain (Piscator 1977).

2. Blood

One study reported 14 and 12 µg Sn/100 g whole blood of "Americans" and Mexican Indians, respectively. Most was in the red cells. Plasma values averaged only 2 µg Sn/100 g. There were large variations from day-to-day and week-to-week primarily attributable to the diet (Stokinger 1981).

C. Excretion

Tin compounds are excreted by the urine (especially inorganic compounds) and bile (especially organotins) (Piscator 1977). Mild alkalosis (due to a vegetable diet) enhances tin excretion in the urine, whereas acidosis decreases it (Stokinger 1981).

IV. EFFECTS

A. Acute and Other Short-Term Exposures

Toxic effects have been seen in people who eat canned food, which has accumulated 250 mg Sn/kg or more; these include nausea, vomiting, and diarrhea. In animal studies, at much higher doses, fatalities are preceded by paralysis and twitching of the limbs, but not frank convulsions.

Organotins are more toxic than inorganic tin compounds. Irritation, even chemical burns, are seen on contact. Healing has been rapid (Piscator 1977; Stokinger 1981). Stokinger (1981) related an unpublished fatal case report involving a woman worker whose clothing was drenched with triphenyltin chloride. Her clothing was not removed immediately nor was her skin adequately cleansed. Consequently, she suffered second- and third-degree burns and died from kidney failure. Respiratory and nervous impairment were not noted.

The only lethal incident associated with tin compounds that has been reported in the literature was the "Stalinon" catastrophe in France, in which about 100 people died from taking an impure, untested drug preparation. Most of the toxicity was associated with the triethyltin contaminant of the alleged diethyltin diiodide. Early symptoms were severe headache, vomiting, vertigo, and visual disturbances. Death occurred from coma, respiratory or cardiac failure, sometimes after meningism, paresis, and convulsions. The lethal mechanism seemed to be cerebral edema (NIOSH 1976; Piscator 1977).

B. Chronic Exposure

Chronic inhalation of tin oxide dust or fumes leads to "stannosis," a benign pneumoconiosis, with normal pulmonary function.

In limited animal studies, both inorganic tin compounds and organotins produced lesions in the brain, kidneys, and liver. The organotins had less effect in the nonneurological systems (Piscator 1977; Stokinger 1981).

C. Biochemistry

1. Effects on Enzymes

No data on inorganic tin compounds. Triethyltin and other trialkyl-tin compounds have been found to inhibit mitochondrial oxidative phosphorylation and brain glucose oxidation (Piscator 1977).

2. Metabolism

No data for inorganic tin compounds. The most toxic organotins are the short-chain trialkyltin compounds. All organotins have the organic moieties removed stepwise to inorganic tin. The toxicity of the tetraalkyltins is apparently due solely to their conversion to trialklytins (Stokinger 1981)

3. Antagonisms and Synergisms

No information.

4. Physiological Requirements

A tin deficiency syndrome has been described in at least one animal species (NAS 1980).

D. Specific Organs and Systems

No data not already cited.

E. Teratogenicity

No teratogenicity was seen in the few studies done (Piscator 1977).

F. Mutagenicity

No credible data. The pesticides triphenyltin acetate and triphenyltin hydroxide were not mutagenic in mouse dominant lethal assays (NIOSH 1976).

G. Carcinogenicity

No carcinogenicity was seen in the few studies found (Piscator 1977; NIOSH 1976).

V. REFERENCES

ACGIH. 1983. TLVs[®] Threshold limit values for chemical substances and physical agents in the work environment with intended changes for 1983-84. Cincinnati, Ohio: American Conference of Governmental Industrial Hygienists.

NAS. 1977. National Academy of Sciences. Drinking water and health. Washington, DC: Printing and Publishing Office, National Academy of Sciences.

NAS. 1980. National Academy of Sciences. Recommended dietary allowances. 9th ed. Washington, DC: Printing and Publishing Office, National Academy of Sciences.

NIOSH. 1976. Natl. Inst. Occupational Safety and Health. Criteria for a recommended standard: occupational exposure to organotin compounds. Washington, DC: U.S. Government Printing Office. DHEW (NIOSH) Publication No. 77-115.

OSHA. 1981. Occupational Health and Safety Administration. Occupational safety and health standards. Subpart 2--Toxic and hazardous substances. Code of federal regulations 29 (Part 1910.1000): 673-679.

Piscator M. 1977. Tin. In: Toxicology of metals - Vol. II. Springfield, VA: National Technical Information Service, pp. 405-426. PB-268 324.

Snyder WS, Cook MJ, Nasset ES, Karhausen LR, Howells GP, Tipton IH. 1975. International Commission on Radiological Protection. Report of the task group on reference man. New York. ICRP Publication 23.

Stokinger HE. 1981. Chapter 29. The metals. In: Patty's industrial hygiene and toxicology. 3rd ed. Volume 2A. Clayton GD, Clayton FE, eds. New York: A Wiley-Interscience Publication. John Wiley and Sons, pp. 1493-2060.

TITANIUM

MAMMALIAN TOXICITY SUMMARY

I. INTRODUCTION

A. Occurrence and Production

Titanium is the ninth most abundant element in the earth's crust (IRPTC Bull. 1983). The major ores are ilmenite (FeO·TiO$_2$) and the less abundant rutile (TiO$_2$). Anatase is another common TiO$_2$ mineral. The metal is produced by chlorinating the oxide and then reducing the titanium tetrachloride (Stokinger 1981).

B. Uses

Most titanium is used, as the dioxide, as a pigment. The metal is used structurally where high strength and light weight overbalance high cost such as in military aircraft and missiles. It is also used in apparatus where inertness is important, as a catalyst (TiCl$_3$), and for a few minor purposes. The very hard TiB$_2$ is used in superalloys and nuclear steels. TiC is also very hard and is used in cutting tools (Stokinger 1981).

C. Chemical and Physical Properties

Titanium, atomic no. 22, is a highly corrosion resistant metal. The major oxidation state is +4, but +2 and +3 compounds are also known. TiCl$_4$ dissolves in cold water but decomposes in hot water to the hydrous oxide and HCl. Titanyl chloride, TiOCl$_2$, probably does not contain the TiO^{2+} ion (Stokinger 1981). Besides the cationic state, anionic titanates are known (IRPTC Bull. 1983).

II. EXPOSURE AND EXPOSURE LIMITS

A. Oral

Titanium levels in food vary widely. Food accounts for practically all of the daily intake, which has been estimated at 300 to 600 µg (Berlin 1977) and 100 to 1,600 µg. Highest titanium concentrations were found in butter and corn oil (Snyder et al. 1975). Snyder et al. (1975) estimate ∿ 2 µg Ti/day from drinking waters.

B. Inhalation

Intake is minor--perhaps 2 or 4 µg/day (Berlin 1977). Ambient air contains up to 1.1 µg Ti/m^3 (Snyder et al. 1975).

Titanium dioxide is classified as a "nuisance particulate" in the workplace (ACGIH 1983).

264

C. Dermal

TiO$_2$ films spread on skin as a protectant from flash burns during World War II were not absorbed and did not cause contact or allergic dermatitis (Stokinger 1981).

D. Total Body Burden and Balance Information

According to Snyder et al. (1975), the titanium balance for the 70-kg reference man in mg/day is: intake 0.850 from food and fluids and 0.001 from air; losses 0.33 in urine, 0.52 in feces, and negligible by other routes. The soft tissue body burden of reference man is estimated as 9.0 mg.

III. TOXICOKINETICS

A. Absorption

Titanium dioxide is only slightly absorbed. To estimate from the amount in urine (10 µg/L), only ~3% of dietary titanium (300 µg/day) is absorbed by humans (Berlin 1977). There is no quantitative information on absorption from the lungs (IRPTC Bull. 1983).

B. Distribution

1. General

The highest single tissue titanium content estimated for reference man is 2.4 mg in lung (Snyder et al. 1975). The general U.S. median for Ti in lung is 220 ppm; for lung in a coal mining area, 380 ppm; and 3,800 ppm, for a coal miner's lung. Striking differences are noted in tissues depending on geographic residence (Stokinger 1981). After lung, kidney and liver usually have highest titanium concentrations (IRPTC Bull. 1983). It can be transported to the brain and the fetus.

2. Blood

Usual titanium concentrations reported in blood are 20 to 70 µg/L (IRPTC Bull. 1983). Snyder et al. (1975) estimate whole blood of reference man contains 0.14 mg Ti with 0.12 mg in plasma and 0.08 mg in red blood cells. Titanium values in the blood of two coal miners were 2.3 and 3 mg/kg (their urine values were 0.11 and 0.49 mg Ti/L compared to 0.0102 mg Ti/L as an average value for Ti in the U.S. population) (Stokinger 1981).

3. Adipose

Snyder et al. (1975) give 0.47 mg (0.031 µg/g) as the adipose tissue content of reference man.

C. Excretion

Data in Snyder et al. (1975) indicate that fecal and urinary excretion are about the same. Presumably, absorbed titanium is primarily excreted in the urine. The average urinary concentration is 10 µg/L. Two reports calculated biological half-lives of 320 and 640 days in humans (IRPTC Bull. 1983).

IV.   EFFECTS

A.   Acute and Other Short-Term Exposures

Titanium tetrachloride liquid is corrosive to skin and eye, producing second and third-degree burns with a long-lasting brown pigment. Its LC_{Lo} in the mouse is 10 mg/m^3 for 2 h.

Death and histopathology in dogs from inhaling $TiCl_4$ were ascribed to the HCl released on hydrolysis. Titanium dioxide and titanium metal are practically inert. The titanium compounds used in dielectrics have extremely low oral toxicity in the rat (Berlin 1977; Stokinger 1981).

Rats drinking water containing 5 ppm Ti K oxalate showed a striking reduction in the numbers of rats surviving to the third generation (IRPTC Bull. 1983).

B.   Chronic Exposure

Titanium dioxide and metal are practically inert. Earlier reports of pulmonary injury by inhaled TiO_2 are now ascribed to contamination with alumina, silica, and other contaminants (Stokinger 1981). A weak fibrogenic response was seen in the lungs of rats inhaling Ti nitride for 6 mo. Ti hydride, boride, and carbide also caused slight fibrosis in rats. Mice drinking water containing 5 ppm Ti for a lifetime showed no adverse effects (IRPTC Bull. 1983).

C.   Biochemistry

1.   Effects on Enzymes

No data.

2.   Metabolism

TiO_2 is not reactive in the body.

3.   Antagonisms and Synergisms

No data.

D.   Specific Organs and Systems

Skin and lung symptoms due to the release of HCl by $TiCl_4$ have been noted above.

E. Teratogenicity

Teratogenic effects have not been reported (IRPTC Bull. 1983).

F. Mutagenicity

No data.

G. Carcinogenicity

Titanium dioxide (anatase) is noncarcinogenic in rats and mice. The organic derivative titanocene, a dicyclopentadiene dichloride, given intraperitoneally, was carcinogenic to C57 Blk mice, a strain resistant to tumors. Various kinds of tumors developed (Berlin 1977; Stokinger 1981). Fibrosarcomas also developed in rats at the injection site of titanium metal (IRPTC Bull. 1983).

V. REFERENCES

ACGIH. 1983. TLVs® Threshold limit values for chemical substances and physical agents in the work environment with intended changes for 1983-84. Cincinnati, Ohio: American Conference of Governmental Industrial Hygienists.

Berlin M. 1977. Titanium. In: Toxicology of metals - Vol. II. Springfield, VA: National Technical Information Service, pp. 85-109, PB-268 324.

IRPTC Bull. 1983. Titanium 6(1):24-26.

Snyder WS, Cook MJ, Nasset ES, Karhausen LR, Howells GP, Tipton IH. 1975. International Commission on Radiological Protection. Report of the task group on reference man. New York. ICRP Publication 23.

Stokinger HE. 1981. Chapter 29. The metals. In: Patty's industrial hygiene and toxicology. 3rd ed. Volume 2A. Clayton GD, Clayton FE, eds. New York: A Wiley-Interscience Publication. John Wiley and Sons, pp. 1493-2060.

TUNGSTEN

MAMMALIAN TOXICITY SUMMARY

I. INTRODUCTION

A. Occurrence and Production

Tungsten is found worldwide in various tungstate minerals. The commerical ores are wolframite, $(Fe, Mn)WO_4$; scheelite, $CaWO_4$; ferberite, $FeWO_4$; and huebnerite, $MnWO_4$. The concentrated tungstate is reduced with hydrogen or carbon to obtain the pure metal (Stokinger 1981).

B. Uses

The largest use of tungsten is conversion to tungsten carbide (WC) for use in cutting and wear-resistent materials. Other uses include alloys, especially for high-strength alloys, incandescent lamp filaments, and compounds used as pigments, catalysts, analytical reagents, and other uses (Stokinger 1981).

C. Chemical and Physical Properties

Tungsten usually has the oxidation state +6, but compounds of +2, +3, +4, and +5 valence are known. Most reactions are similar to those of molybdenum (Stokinger 1981).

II. EXPOSURE AND EXPOSURE LIMITS

A. Oral

One study estimated dietary intake of 8 to 13 µg/day (Kazantzis 1977).

B. Inhalation

No data on general exposures are available.

The threshold limit value (TLV) as a time-weighted average in workroom air for an 8-h day is 1 mg W/m^3 for soluble tungsten compounds; the short-term exposure limit (STEL) is 3 mg/m^3 for 15 min. For insoluble tungsten compounds, the TLV is 5 mg W/m^3 and the STEL, 10 mg W/m^3 for 15 min (ACGIH 1983).

C. Dermal

No data are available.

D. Total Body Burden and Balance Information

No data are available.

III. TOXICOKINETICS

A. Absorption

In single studies, oral tungstate was well absorbed by rats and inhaled tungstic acid (WO_3) was poorly absorbed (Stokinger 1981).

B. Distribution

1. General

In the oral study, bone and spleen had the highest concentrations. In the inhalation study, lungs and kidneys predominated (Stokinger 1981).

2. Blood

Normal tungsten values was not found. In a Russian study, 88 of 178 workers exposed to 0.75 to 6.1 mg W/m³ and 0.6 to 3.2 mg Co/m³ complained of dyspnea, coughing, pounding of the heart, headache, dizziness, nausea, loss of appetite, and impaired sense of smell. Although not detected in the blood of 11 workers, mean blood tungsten concentrations of 45 workers were 0.8 to 1.1 mg% (NIOSH 1977).

C. Excretion

Urine seems to be the main excretory route (Kazantzis 1977). In 40 of the 188 Russian workers mentioned above, urine tungsten concentrations were 0.6 to 1.1 mg/L, but it was not detected in the urine of seven workers (NIOSH 1977).

IV. EFFECTS

A. Acute and Other Short-Term Exposures

There is no evidence of acute toxicity of tungsten and its compounds in humans. Soluble compounds are generally more toxic by every route than are the metal and insoluble compounds. Lethal and near-lethal doses produce nervous prostration, diarrhea, coma, and death due to respiratory paralysis. Repeated doses not immediately fatal cause anorexia, colic, incoordinated movements, trembling, and weight loss (Stokinger 1981).

B. Chronic Exposure

Worker exposure, primarily to WC in grinding wheels, produces what is called "hard metal disease." The main symptoms are cough, dyspnea, and wheezing with minor radiological abnormalities. Hypersensitivity, asthma and marked radiological abnormalities are seen in some. This progresses to diffuse interstitial pulmonary fibrosis, which can cause (or contribute) to deaths from cor pulmonale, cardiac failure, and emphysema. The disease is primarily attributable to the cobalt content (Kazantzis 1977; NIOSH 1977; Stokinger 1981).

C. Biochemistry

1. Effects on Enzymes

Tungstate has been shown to inhibit the molybdate-dependent enzymes xanthine dehydrogenase, sulfite oxidase and aldehyde oxidase, and nitrate reductase. It activates brain glutaminase and inactivates liver glutaminase (Stokinger 1981).

2. Metabolism

No information.

3. Antagonisms and Synergisms

No information.

D. Specific Organs and Systems

Tungsten is basically inert. See IV.B above.

E. Teratogenicity

No information.

F. Mutagenicity

No information.

G. Carcinogenicity

No information.

V. REFERENCES

ACGIH. 1983. TLVs® Threshold limit values for chemical substances and physical agents in the work environment with intended changes for 1983-84. Cincinnati, Ohio: American Conference of Governmental Industrial Hygienists.

Kazantzis G. 1977. Tungsten. In: Toxicology of metals - Vol. II. Springfield, VA: National Technical Information Service, pp. 442-453. PB-268 324.

NIOSH.   1977.   Natl. Inst. Occupational Safety and Health.   Criteria for a recommended standard: occupational exposure to tungsten and cemented tungsten carbide.   Washington, DC: U.S. Government Printing Office.   DHEW (NIOSH) Publication No. 77-127.

Stokinger HE.   1981.   Chapter 29.   The metals.   In: Patty's industrial hygiene and toxicology.   3rd ed.   Volume 2A.   Clayton GD, Clayton FE, eds. New York: A Wiley-Interscience Publication.   John Wiley and Sons, pp. 1493-2060.

MAMMALIAN TOXICITY SUMMARY

I.   INTRODUCTON

A.   Occurrence and Production

Uranium is a moderately common element, found in most rocks, with greatest concentrations (8 ppm) in acidic igneous rocks (granites). Several complex minerals are of commercial importance, including carnotite $(K_2O \cdot 2U_2O_3 \cdot V_2O_5 \cdot 3H_2O)$, pitchblende (a complex of $UO_3 \cdot UO_2$, PbO, etc.), and tobernite $(Cu(UO_2)_2P_2O_8 \cdot 12H_2O)$. The latest recovery process involves extraction by the bacterium _Thiobacillus ferroxidans_, followed by electrolytic reduction, and precipitation as "green cake," the tetrafluoride, which is further converted to the metal, oxide, hexafluoride, etc. (Stokinger 1981).

B.   Uses

Most uranium is enriched with ^{235}U or irradiated to ^{239}Pu for use in nuclear piles or in weapons. Small quantities are used in ceramics, glass, and for various chemicals (Stokinger 1981).

C.   Chemistry

All naturally occurring uranium isotopes are radioactive. The main compounds have +4 and +6 oxidation states, with +3 and +5 also known but unstable in aqueous media. U^{3+} oxidizes readily. U^{5+} disproportionates to U^{4+} and U^{6+}. UO_2^{2+} is the form of U(VI) that exists in acid solution and in physiological systems (Stokinger 1981). U(IV) and U(VI) complex with carbonate and proteins in the body (Berlin and Rudell 1977).

II.   EXPOSURE AND EXPOSURE LIMITS

A.   Oral

Exposure estimates are widely divergent. Daily intakes in urban areas of various countries have been estimated as 1.0 to 1.5 µg/day. Table salt, vegetables, and cereals contribute the most to the dietary intake (Berlin and Rudell 1977). In districts overlying uranium deposits, drinking water may be an important source (e.g., 15 µg/L in a Yugoslavian uranium mine area). In New York, the content in drinking water is ~ 0.032 µg/L (Snyder et al. 1975).

B.   Inhalation

Ambient air in New York contains 0.4 mg U/m^3 (Snyder et al. 1975).

The threshold limit value as a time-weighted average in workroom air for an 8-h day is 0.2 mg U/m^3 for natural uranium (unenriched) and its compounds; the short-term exposure limit is 0.6 mg/m³ for 15 min (ACGIH 1983). The OSHA permissible exposure limit is 0.05 mg/m³ for soluble uranium compounds and 0.25 for insoluble compounds (OSHA 1981).

C.   Dermal

No data.

D.   Total Body Burden and Balance Information

The estimated body burden is 100 to 125 µg/man (Stokinger 1981). According to Snyder et al. (1975), the uranium balance for 70-kg reference man in µg/day is: Intake 1.9 from food and fluids and 0.002 from air; losses 0.05 to 0.5 in urine, 1.4 to 1.8 in feces, and 0.02 in hair.

III.   TOXICOKINETICS

A.   Absorption

Absorption is primarily a function of solubility of +6 compounds, which are usually absorbed as UO_2^{+2}. Tetravalent compounds are not absorbed (Stokinger 1981). In general, only small amounts of ingested or inhaled uranium are actually absorbed (0.5 to 5%, for instance) except that alveolar deposits are eventually completely absorbed. Some water-soluble compounds are absorbed through the skin (Berlin and Rudell 1977).

B.   Distribution

1.   General

It is widely distributed, with large concentrations in the kidneys and very large deposits in the skeleton (Berlin and Rudell 1977; Stokinger 1981).

2.   Blood

Uranium is transported in the blood as carbonate complexes in the plasma (47%), plasma protein complexes (32%), and in red blood cells (20%). Clearance from the bloodstream is rapid (Berlin and Rudell 1977).

C.   Excretion

Excretion of absorbed uranium is through the urine and part of the dose is very slowly excreted and part, rapidly. Measurable uranium in urine can be found 1.5 yr after a single, large dose. The biological half-life of UO_2 in the lungs is ~ 15 mo for monkeys and dogs (Berlin and Rudell 1977; Stokinger 1981). Overexposure to soluble uranium compounds may be monitored by the urinary excretion of uranium. Levels in the urine of nonoccupationally exposed persons are 0.03 to 0.3 µg/L. At the end of the shift, worker urine concentrations should not exceed 250 µg/L to avoid renal damage (Lauwerys 1983).

IV. EFFECTS

A.   Acute and Other Short-Term Exposures

Oral toxicity of soluble compounds is rather low, but the soluble compounds are highly toxic by inhalation or by placement in the conjunctival sac. UF_4 was the most toxic by inhalation and UO_2 and U_3O_8, the least toxic (Stokinger 1981).

Short-term exposures to soluble uranium compounds exhibit the chemical toxicity of uranium. Large amounts of uranium rapidly pass through the kidney and damage the proximal convoluted tubules. The new epithelial lining that forms after a nonlethal dose differs morphologically from normal tubular epithelium and may contribute to the tolerance to uranium developed by animals. However, after cessation of exposure, the epithelium returns to its normal appearance. Damage to the glomeruli (primarily to the basement membrane of the capillaries) is evident from proteinuria, impaired diodrast and PAH clearance, and increased clearance of amino acids and glucose. Earliest signs of tubular damage are albuminuria and increased urinary catalase. As little as 0.02 mg uranyl nitrate per kilogram in rabbits caused the presence of catalase in the urine. Acute toxic doses of soluble uranium salts also damage brain structures (Berlin and Rudell 1977; Stokinger 1981).

B.   Chronic Exposure

Enriched uranium is a highly significant radiation hazard. Miners develop a high incidence of respiratory disease (including cancers), primarily from radon daughters. Absorbed uranium, deposited in the bone, would be highly toxic to the hematopoiesis in the marrow (Stokinger 1981).

Soluble uranium salts were tolerated for 1 yr in the diet of dogs at 0.2 mg/kg/day. The tolerable level for UO_2 was 10 g/kg/day. In rats fed uranyl nitrate hexahydrate in the diet for 2 yr, 0.1% and 0.5% caused barely detectable growth depression in male and female rats, respectively. Even 2% did not alter the lifespan (Stokinger 1981).

C.   Biochemistry

1.   Effects on Enzymes

Not well known, although many uranium-protein complexes exist (Stokinger 1981).

2.   Metabolism

Hexavalent uranium is reduced to the tetravalent form intracellularly; it is probable that the reverse process also occurs. Soluble uranium will immediately complex the bicarbonate, citrate, malate, lactate, and other bases in the body. The uranyl cation will be bound to phosphate-containing molecules and to carboxyl and/or hydroxyl groups of proteins, nucleotides, and bone tissue. Uranium is bound in the hydroxyapatite complex of bone, substituting for calcium (Berlin and Rudell 1977).

3. Antagonisms and Synergisms

No data.

D. Specific Organs and Systems

See remarks on the kidney in part III.B.

E. Teratogenicity

Except for radioactivity, none known.

F. Mutagenicity

Except for radioactivity, none known.

G. Carcinogenicity

Except for radioactivity, none known.

V. REFERENCES

ACGIH. 1983. TLVs® Threshold limit values for chemical substances and physical agents in the work environment with intended changes for 1983-84. Cincinnati, Ohio: American Conference of Governmental Industrial Hygienists.

Berlin M, Rudell B. 1977. Uranium. In: Toxicology of metals-Volume II. Springfield, Virginia: National Technical Information Service, pp. 454-472. PB-268 324.

Lauwerys RR. 1983. Chapter II. Biological monitoring of exposure to inorganic and organometallic substances. In: Industrial chemical exposure: Guidelines for biological monitoring. Davis CA: Biomedical Publications, pp. 950.

OSHA. 1981. Occupational Safety and Health Admin. Occupational safety and health standards. Subpart 2--Toxic and hazardous substances. Code of federal regulations 29 (Part 1910.1000) pp. 673679.

Snyder WS, Cook MJ, Nasset ES, Karhausen LR, Howells GP, Tipton IH. 1975. International Commission on Radiological Protection. Report of the task group on reference man. New York. ICRP Publication 23.

Stokinger HE. 1981. Chapter 29. The metals. In: Patty's industrial hygiene and toxicology, 3rd ed., volume IIA, Toxicology. Clayton GD, Clayton FE, eds. New York: Wiley-Interscience, John Wiley & Sons, pp. 1493-2060.

MAMMALIAN TOXICITY SUMMARY

I. INTRODUCTION

A. Occurrence and Production

The earth's crust contains an average of 150 ppm vanadium in the form of relatively insoluble salts. In nature, vanadium generally occurs in the trivalent state. Vanadium is extracted in the United Sates from carnotit [$K(UO_2)_2(VO_4)_2$], phosphate rock deposits, titanium-bearing magnetite associated with vanadium; and vanadium-bearing clays. Roasting vanadium-bearing material with 6 to 10% NaCl at 850°C converts it to water-soluble sodium metavanadate ($NaVO_3$). Leaching followed by acid precipitation gives the commerical V_2O_5. Ferrovanadium is produced by smelting scrap steel and $NaVO_3$ in electric arc furnaces (NAS 1974).

B. Uses

Ferrovanadium is used to produce construction steels and tool-and-die steels. Vanadium is also used in high-strength titanium and aluminum alloys. Vanadium compounds [$VOCl_2$, VCl_4, and V tris(acetylacetonate)] are used as polymerization catalysts for ethylene-propylene copolymers and synthetic rubbers. Other vanadium catalysts are used in sulfuric acid synthesis (SO_2 to SO_3), oxidation of organic compounds, and aniline black production. Vanadium compounds are also used as mordants in dyeing and printing cotton and for fixing aniline black on silk. Ammonium metavanadate, NH_4VO_3, may be present in quick-drying inks (NAS 1974).

C. Physical and Chemical Properties

Vanadium, atomic No. 23, is a gray, relatively soft metal, which improves the strength and other properties of steel at concentrations as low as 0.02%. Vanadium exhibits oxidation states of +2, +3, +4, and +5 in its compounds. VO and V_2O_3 are basic, VO_2 is basic or weakly acidic, and V_2O_5 is amphoteric. The V(II) compounds are powerful reducing agents and V(III) are reducing agents as well. VO^{2+} is termed the vanadyl radical; but to avoid confusion, $VOCl_2$ is called vanadium oxydichloride and $VOCl_3$ is vanadium oxytrichloride.

V_2O_5 is very slightly water soluble and V_2O_3 is slightly soluble in cold water; V_2O_4 is insoluble. $VOCl_2$ and $VOCl_3$ dissolve in water with decomposition. $VOSO_4$, NH_4VO_3, and $NaVO_3$ are highly to extremely water soluble. Vanadyl esters--$VO(OR)_3$ or $V(OR)_3$ where R is ethyl, propyl, or butyl--and water-soluble acetylacetonates--$V(C_5H_7O_3)_3$ and $VO(C_5H_7O_3)_2$--have replaced NH_4VO_3 and $NaVO_3$ for use as catalysts, paint and ink driers, and mordants and fixers (Stokinger 1981).

II. EXPOSURE AND EXPOSURE LIMITS

A. Oral

Vanadium concentrations in plant and animal foods are probably on the order of a few parts per billion wet weight. Seafoods, calf liver, and gelatin contain 2.4 to 44 ppb as determined by neutron activation analysis. Older analytic data are questionable (NAS 1974). Vanadium was found in the finished water of the 100 largest U.S. cities at a median concentration of < 4.3 µg/L with a maximum level of 70 µg/L (Durfor and Becker 1964: cited by NAS 1977). In 380 finished waters vanadium was detected in 3.4% with a mean concentration of 46.1 µg/L and a range of 14 to 222 µg/L (Kopp 1970; cited by NAS 1977).

B. Inhalation

Sources of vanadium in ambient air are combustion of coal, crude oils, and undesulfurized heavy fuel oils. Twice as much V appears in the air during the heating season as during spring and summer. Ambient air concentrations in cities of the Northeast United States have averaged up to 1,320 ng V/m^3. Average concentrations in the West and Midwest range up to 22 ng V/m^3 (NAS 1974).

The threshold limit value as a time-weighted average in workroom air for an 8-h day is 0.05 mg V_2O_5/m^3 for vanadium respirable dust and fume (ACGIH 1983). The OSHA permissible exposure limit is 0.1 mg V_2O_5/m^3 for vanadium dust, 0.5 mg V_2O_5/m^3 for vanadium fume (OSHA 1981).

C. Dermal

NAS (1974) stated that environmental exposure for vanadium from this route is apparently of minor importance.

D. Total Body Burden and Balance Information

According to Snyder et al. (1975), the vanadium balance for 70-kg reference man (mg/day) is: Intake 2.0 from food and fluids and 0.2×10^{-3} from air; losses 0.015 from urine and 2.0 feces. Snyder et al (1975) estimate a soft tissue body burden of < 18 mg.

III. TOXICOKINETICS

A. Absorption

The oxidation state of vanadium does not appear to influence absorption. Stokinger (1981) reviews many feeding studies but does not quantitate absorption from this route. No information is given from which the degree of absorption from the lungs can be quantitated. Vanadium was absorbed through rabbit skin from a 20% solution of $NaVO_3$ (Stokinger 1981).

B. Distribution

1. General

Highest vanadium concentrations are found in the lungs of humans. Lower small and large intestines, omentum, and skin also accumulate vanadium. Tissue concentrations in lung increase with age (NAS 1974). Stokinger (1981) reviewed animal tissue distribution after ingestion and parenteral administra-tion. Retention is high in bone in the rat because of the constant metabolic turnover of bone in this species.

2. Blood

Normal whole blood vanadium concentrations are < 1 µg/L (Lauwerys 1983).

Kiviluoto et al. (1979), using atomic absorption spectrophotometry with adequate to good quality control, reported that vanadium concentrations in neither blood serum nor urine correlated with the degree of low-level inhalation exposure (0.01 to 0.04 mg V/m^3) in vanadium smelter workers. However, while the mean serum concentration was 0.22 ± 0.14 µmol/L in the vanadium-exposed group and 0.26 ± 0.17 µmol/L in the urine, vanadium was not detected in the urine of the control group.

Serum vanadium concentration in moderately exposed workers fell from 393 ± 223 nmol/L at the start of their vacation to 225 ± 83 nmol/L 16 days later, but the difference was not statistically significant. Serum vanadium correlated with creatinine-adjusted urinary vanadium concentrations at the start of the holiday (Kiviluoto et al. 1981a). Kiviluoto et al. (1981b) were also unable to show an association between the results of hematologic and chemical laboratory tests and exposure to 0.2 to 0.5 mg V/m^3. The minor differences of the results of some chemical laboratory tests from the results of the controls were not clinically significant. There were no hematologic abnormalities in workers exposed to 0.01 to 0.04 mg V/m^3 in the factory air. The mean serum values in the exposed workers was 0.22 ± 0.14 µmol/L as stated above.

Maroni et al. (1983) found that urinary vanadium but not blood vanadium concentrations of workers involved in cleaning an oil-burning boiler was correlated with exposure. (The absolute degree of exposure was unknown since the workers were wearing respiratory protection.) Values for unexposed persons (including the maintenance operators) ranged from 0.94 ± 0.65 to 2.83 ± 2.20 µg/L (determined by neutron activation analysis, but no standard reference materials were used for quality control). After exposure, blood values were 0.83 ± 0.29 to 5.45 ± 3.61 µg/L. At the same time, urinary vanadium values had increased several fold from the baseline values.

In unexposed humans, 95% of the blood vanadium is in the trans-ferrin of the plasma. In vitro experiments show that unbound vanadium con-centrations in blood must be > 100 times higher than that of unexposed sub-jects to significantly decrease human serum alkaline phosphatase activity (Marafante et al. 1981). Most absorbed vanadium in workers with moderate long-term exposure is excreted within 1 day after cessation of exposure (Kiviluoto et al. 1981a).

Workers excreted 73 ± 50 nmol V/mmol creatinine in the urine at the beginning of their vacation; after 3 days, the urine concentration of 46 ± 24 nmol/mmol creatinine was not different from the concentration after 16 days and was close to (but significantly higher than) that of their unexposed family members (32 ± 17 nmol/mmol creatinine).

C. Excretion

Lung clearance of $^{48}VOCl_2$ was rapid; after 1 day > 50% had been removed. The clearance rate slows thereafter; 3% of the ^{48}V lung burden remained after 9 wk. Absorbed vanadium is predominantly excreted in the urine (Stokinger 1981). Less than 1% of ingested vanadium is absorbed (Lauwerys 1983).

IV. EFFECTS

A. Acute and Other Short-Term Exposures

V salts exhibit high parenteral toxicity, intermediate inhalation toxicity, and low oral toxicity. The oral LD_{50} of V_2O_3 in mice is 130 mg V/kg; that of VCl_3 is 23 mg V/kg. The vanadium alcoholates are extremely irritating to the mucous membranes and can cause circulatory problems.

Eight-hour exposures of humans to 0.1 to 1 mg V_2O_5/m^3 produced mucus and/or coughing that persisted 4 to 10 days without systemic complaints. Hypersensitivity appeared in two men subsequently exposed for 5 min to a dense cloud of V_2O_5 dust, who developed marked coughing with sputum and rales and expiratory wheezing.

Industrial exposures are generally described as acute episodes with relapses and sometimes chronic coughing and chronic bronchitis as sequelae. Symptoms, generally upper and lower respiratory tract irritation, do not appear until after repeated exposures to vanadium compounds for a few days or a week, which may indicate development of delayed hypersensitivity. Conjunctiva are irritated and eczema may develop. Respiratory symptoms are disabling (Stokinger 1981).

B. Chronic Exposure

Sequelae of acute vanadium intoxications may include chronic respiratory symptoms, but pneumoconiosis, fibrosis, and emphysema do not develop (NAS 1974).

Epidemiology investigations have correlated concentrations of vanadium and other metals in community air with disease mortality indexes. V, Cd, Zn, Sn, and Ni in the air of 25 United States communities correlated fairly well with "diseases of the heart," nephritis, and "arteriosclerotic heart." In Great Britain, V, As, and Zn in the air of 23 localities showed weak association with lung cancer; V was strongly associated with bronchitis in males; V and Be were associated with pneumonia; and V, Be, and Mo correlated with other cancers, except of stomach, in males. Vanadium was the suspected, but never proved, etiological agent of the upper respiratory symptoms called Yokahama asthma that developed in U.S. servicemen and their families stationed in Japan in the early 1960's. Imported fuel oils from the Middle East contained high vanadium levels (NAS 1974). Workers exposed to vanadium have lowered cystine content in their fingernails, which can be used as a diagnostic test (NAS 1974). Tolerance to the acute respiratory effects develops after repeated exposure to small doses (Stokinger 1981).

C. Biochemistry

1. Effects on Enzymes

Vanadium reduces the concentration of coenzyme A in the liver and inhibits the enzyme squalene synthetase. In young animals, vanadium depresses conversion of acetate to cholesterol by activating mitochondrial acetoacetylcoenzyme A deacylase; in older animals, the enzyme was inhibited by vanadium. Vanadium decreases coenzyme A in mitochondria and may activate or inactivate tissue monoamine oxidase depending on its concentration. Succinoxidase activity is reduced in the presence of vanadium; this reduction would lead to reduced ATP synthesis (NAS 1974).

2. Metabolism

Vanadium affects several metabolic processes. Destruction of cystine and cysteine is increased in the presence of elevated levels of vanadium. The decrease in coenzyme A synthesis is due to the decrease of cystine concentration since thioethanolamine, a reduction product from cystine, is one of the compounds involved in coenzyme A synthesis. Metabolic processes that use acetate as a starting material, such as cholesterol synthesis, will be inhibited. Serum cholesterol lowering has been observed in young men given diammonium oxytartratovanadate for 6 weeks and in workers exposed to high-vanadium dusts although other reports do not confirm this. Vanadium has a hematopoietic effect (increase in hemoglobin in the blood), but the mechanism is not known (NAS 1974).

3. Antagonisms and Synergisms

The depression of coenzyme Q synthesis by vanadium can be counteracted by simultaneous administration of adenosine triphosphate (ATP), coenzyme A, and cysteine (NAS 1974). Ascorbic acid given to rats, mice, and dogs protects against and is antidotal to vanadium poisoning (Arena 1973; Stokinger 1981).

4. Physiological Requirements

Vanadium has been shown to be essential in rats and chicks, but its role is not known. It will probably be found essential for other animals (NAS 1974).

D. Specific Organs and Systems

1. Respiratory System

Irritation of the human respiratory tract occurs with concentrations of V_2O_5 and NH_4VO_3 as low as 0.04 mg V/m^3 and 5-min exposure to 0.6 mg V/m^3 as V_2O_5 is sufficient to produce coughing and rales. Acute exposure to the oxides and vanadates cause wheezing, rales, rhonchi, and chest pain. Ferrovanadium exposure does not cause such severe symptoms. Chest pain, asthma-like bronchitis, and emphysema (not conclusive) have been reported in workers exposed to vanadium compounds for $\geq$ 6 mo (NIOSH 1977).

2. Skin and Eyes

Vanadium compounds except ferrovanadium irritate the skin and eyes. A sensation of heat or itching sometimes with rashes or eczema occurs at air concentrations $\geq$ 0.03 mg V/m^3. Conjunctivitis or a burning sensation in the eyes occurs at $\geq$ 0.018 mg V/m^3. No permanent eye damage from vanadium exposure has been reported (NIOSH 1977).

3. Tongue and Oral Mucosa

A greenish-black discoloration of the tongue and oral mucosa may occur at concentrations $\geq$ 0.08 mg V/m^3, sometimes with a salty or metallic taste, and disappears within 2 to 3 days after cessation of exposure (NIOSH 1977).

E. Teratogenicity

NIOSH (1977) found no teratogenic information on vanadium compounds

F. Mutagenicity

NIOSH (1977) found no references to vanadium mutagenicity in its literature review.

G. Carcinogenicity

Vanadyl sulfate at 198 mg/kg body wt./y administered to Charles River white swiss mice for their lifespan did not affect the tumor incidence compared to that of the controls (Kanasawa and Schroeder 1967; cited by NIOSH 1977).

V. REFERENCES

ACGIH. 1983. TLVs[®] Threshold limit values for chemical substances and physical agents in the work environment with intended changes for 1983-84. Cincinnati, Ohio: American Conference of Governmental Industrial Hygienists.

Arena JM. 1973. Poisoning: toxicology--symptoms--treatments. Springfield, IL: Charles C. Thomas Publisher.

Kiviluoto M, Pyy L, Pakarinen A. 1979. Serum and urinary vanadium of vanadium-exposed workers. Scand J Work Environ Health 5(4):362-367.

Kiviluoto M, Pyy L, Pakarinen A. 1981a. Serum and urinary vanadium of workers processing vanadium pentoxide. Int Arch Occup Environ Health 48:251-256.

Kiviluoto M, Pyy L, Pakarinen A. 1981b. Clinical laboratory results of vanadium-exposed workers. Arch Environ Health 36(3):109-113.

Lauwerys RR. 1983. Chapter II. Biological monitoring of exposure to inorganic and organometallic substances. In: Industrial chemical exposure: Guidelines for biological monitoring. Davis, CA: Biomedical Publications, pp. 9-50.

Marafante R, Pietra R, Rade-Edel J, Sabbioni E. 1981. Vanadium: metabolic fate and identification of V-binding components in laboratory animals and human blood. In: Heavy met environ, Int Conf, 3rd. Edinburgh, United Kingdom: CEP Consultants Ltd., pp. 498-501.

Maroni M, Colombi A, Buratti M, Calzaferri G, Foa V, Sabbioni E, Pietra R. 1983. Urinary elimination of vanadium in boiler cleaners. Heavy Met Environ, Int Conf, 4th, 1:66-69.

NAS. 1974. Natl. Academy of Sciences. Vanadium. Washington, DC: Printing and Publishing Office, 117 pp.

NIOSH. 1977. Natl. Inst. Occupational Safety and Health. Criteria for a recommended standard: occupational exposure to vanadium. Washington, DC: U.S. Government Printing Office, DHEW (NIOSH) Publication No. 77-222.

OSHA. 1981. Occupational Safety and Health Admin. Occupational safety and health standards. Subpart 2--Toxic and hazardous substances. Code of federal regulations 29 (Part 1910.1000) pp. 673-679.

Snyder WS, Cook MJ, Nasset ES, Karhausen LR, Howells GP, Tipton IH. 1975. International Commission on Radiological Protection. Report of the task group on reference man. New York. ICRP Publication 23.

Stokinger HE. 1981. Chapter 29. The metals. In: Patty's industrial hygiene and toxicology. 3rd ed. Volume 2A. Clayton GD, Clayton FE, eds. New York: A Wiley-Interscience Publication. John Wiley and Sons, pp. 1493-2060.

MAMMALIAN TOXICITY SUMMARY

I. INTRODUCTION

Browning (1969) gives brief separate summaries for cerium (Ce) and lanthanum (La). Stokinger (1981) discusses the group together because of their chemical and toxicological similarities. Haley (1965) published an earlier review with 96 references.

A. Occurrence and Production

The crustal abundance of the naturally occurring rare earths ranges from 0.2 ppm for thulium (Tm) to 46.1 ppm for cerium (Haley 1965). Rare earth metals are recovered from ores containing massive monazite and monazite sand (Ce group metal phosphates); bastnasite and related Ce group fluorocarbonates; and the yttrium group minerals gadolinite (Y, Ce, Cr, Be, Fe silicate), euxenite (contains Y, Ce, Er, Nb, Ti, U), and xenotime (YPO$_4$ with Th and some Ce subgroup metals). Lanthanides are recovered by precipitation of the oxalates from a sulfuric acid solution of the ore (thorium is first removed as a pyrophosphate precipitate). Boiling with NaOH solution gives granular hydroxides. Calcining gives the oxides. To separate cerium, it is oxidized to the tetravalent state; no other lanthanides exist in this state. Commercial lanthanide salts are mixtures containing the rare earth metals in about the same ratio as in the ore. Fractional crystallization or ion-exchange methods are used to separate the elements (Stokinger 1981). Browning (1969) described somewhat different industrial methods. Rhone-Poulenc Chemical Company operates a separation plant in Freeport, Texas (Chem Eng News 1981).

B. Uses

Lanthanide compounds are used in carbon-arc lighting for studio lighting, theater projection, and searchlights. Mixed lanthanide metals (misch metal*) and cerium metal are used in lighter flints, magnesium alloys, and some iron alloys. Lanthanide salts are sometimes used instead of the metals in alloy production. Other salts are used in the coloring, decolorizing, and polishing of glass. About a quarter of consumption is for miscellaneous uses (Stokinger 1981). Uses for rare earths listed in a 1981 Rhone-Poulenc advertisement (Chem Eng News 1981) included phosphors for color televisions, medical x-rays, fluorescent bulbs, coloring and ultraviolet filtering agents in glass and ceramics, catalytic uses, small high-strength magnets, surgery and instrument gauges, lasers, and computer bubble memories. Ce oxalate has been used to treat the nausea and vomiting of pregnancy (Browning 1969).

---

* Electrolysis of the fused mixed chlorides from monazite sand gives the alloy misch metal (Browning 1969).

C. Physical and Chemical Properties

The chemistry and toxicity of the rare earths are similar enough to warrant their discussion as a group. Melting points range from 799°C for cerium (atomic no. 58) to 2822°C for europium (atomic no. 63). The common oxidation state is III. Some also form compounds with oxidation state IV ($_{58}$Ce, $_{59}$Pr, $_{65}$Tb) or II ($_{62}$Sm, $_{63}$Eu, $_{70}$Yb). All form metal(III) oxides as ignition products in air. Promethium (Pm) does not occur naturally. All of these elements alloy with most other metals. Salts precipitate at physiological pH and form insoluble complexes with nucleic acids (Stokinger 1981).

II. EXPOSURE AND EXPOSURE LIMITS

A. Oral

No information was found in the secondary sources consulted.

B. Inhalation

Several of the rare earths are determined in multielement analyses of ambient air. No systematic collection of such literature could be made for the present toxicity summary. ACGIH (1983) recommends a threshold limit value of 1 mg/m³ for yttrium in workplace air (8-h time weighted average). The short-term exposure limit is 3 mg/m³/15 min. The OSHA permissible exposure limit is 1 mg/m³ (OSHA 1981).

C. Dermal

No information.

D. Total Body Burden and Balance Information

No information was found in the secondary sources consulted.

III. TOXICOKINETICS

A. Absorption

Less than 1% of administered oral doses of the radioisotopes [144]Ce, [152,154]Eu, [160]Tb, and [170]Tm were absorbed by rats within 4 days. Most lanthanides are absorbed from parenteral injection sites within 4 days except [140]La, only 87.5% of which was absorbed in that time due to its low solubility at body pH (Stokinger 1981).

B. Distribution

1. General

All lanthanides deposit in the bone in amounts ranging from 62% of the dose administered for Gd to 90% for Lu although the initial deposition is nearly equal for Eu and Gd in liver and bone. For the lighter lanthanides, the initial major deposition site is the liver (45% for Sm to 67% for La). However, kidneys originally contain the highest lanthanide concentrations, but the levels are very low after 8 mo. Intravenous doses of lanthanide halides or nitrates form colloids, which accumulate in the reticuloendothelial system (Stokinger 1981). Bone uptake of La from i.v. doses of $LaCl_3$ was highest at lowest doses. Early distribution of La includes spleen and muscle as well as liver and bone (Browning 1969).

2. Blood

In the rat experiments described in III.A., concentrations in the blood at 1 day after injection were < 0.02%/mL and decreased steadily thereafter (Stokinger 1981). Approximately 50% of intravenously administered La is cleared from the bloodstream within 1 h (Browning 1969).

C. Excretion

After accumulation in the liver, the light lanthanides are excreted in the feces. Urinary excretion predominates for the heavier lanthanides. The middle lanthanides are excreted about equally in the urine and feces (Stokinger 1981). The rate is slow. Only 80% of an i.v. dose of $LaCl_3$ (tracer activity 15 to 20 µCi) was excreted within 96 h (Browning 1969).

IV. EFFECTS

A. Acute and Other Short-Term Exposures

No definitive evidence of occupational poisoning has been reported. Indefinite symptoms of headache and nausea may have been due to lithographic industry worker exposure to La_2O dust and fume from cored arc-light carbons (Stokinger 1981).

Total intravenous doses of 250 to 500 mg Nd given to prolong blood coagulation time in humans were without ill effects. The anticoagulant effect lasted for 6 h. However, single and repeated daily doses of Nd, La, and Ce at 3 to 12.5 mg/kg in 18 patients produced chills, fever, headache, muscle pains, abdominal cramps, hemoglobinemia, and hemoglobinuria. The anticoagulant effect and Nd concentrations in the blood persisted for 8 h. Increases of $\geq$ 9% hemoglobin in the blood serum indicate overexposure to lanthanides. Injection of $LaCl_3$ has caused local thrombophlebitis in humans and dogs (Stokinger 1981).

Stokinger (1981) summarized the acute toxicity experiments (with some discrepancies with Haley's values). Haley (1965) more clearly presents the information in tables. The intraperitoneal LD_{50} values for all the nitrates (except Pm) for mice and rats range from 210 to 480 mg/kg. Intravenous LD_{50}'s ranged from 49.6 to 77.2 mg/kg in the male rat and from 4.3 to 35.8 mg/kg in the female rat for Ce, Pr, Nd, Sm, and Er. Because of poor gastrointestinal absorption, oral LD_{50}'s of the nitrates are about

(Stokinger 1981). Other metabolic effects (on enzymes, the heart, and the liver) are described in IV.C.1, IV.C.3, and IV.D.1.

3. Antagonisms and Synergisms

La^{3+} antagonizes Ca^{2+} binding in heart muscle, which inactivates the contractile chemotactic process. The blood anticoagulant activity of the lanthanides has been studied for the prevention of thrombosis (Stokinger 1981). The anticoagulant effect of the rare earth salts is counteracted by vitamin K (Haley 1965).

D. Specific Organs and Systems

1. Liver

Repeated intravenous injections of cerium in high doses in rats caused fatty infiltration of the liver, which was more pronounced in females. Neutral fat esters increased, but total cholesterol and phospholipid concentrations were not changed. Mitochondria were enlarged. Female rats especially showed changes in the ground substance and cristae and lowering of the blood glucose for the first 3 or 4 days. Both sexes showed changes in the endoplasmic reticulum: dilatation of the cisternae and a loss of ribosome after treatment with each of the lanthanides (Stokinger 1981) (Browning 1969). The fatty liver is reversible without therapy (Haley 1965).

2. Skin and Eye

Many lanthanide salts irritate or damage the eyes and abraded skin. Er irritates even intact rabbit skin. Intradermal injections produced foreign body reactions (granulomas) without resorption of the crystalline deposit within a 45-day observation period (Stokinger 1981). The rare earths are highly irritating to the conjunctiva but cause opacification of the cornea when denuded only after several hours or days (Haley 1965).

3. Lung

Edema and pleural effusion were observed in rats dosed with soluble cerium compounds (Browning 1969).

4. Heart and Other Muscles

Rare earths produced a negative inotropic effect prior to paralysis of the heart in vitro. In vivo intravenous doses cause ECG changes and cardiovascular collapse (Haley 1965). Several of the rare earth salts decrease the tonus and produce loss of contractility of isolated ileum and uterus due to a nonspecific antispasmodic effect against acetylcholine. Ce nitrate had a curariform effect on striated muscle.

5. Blood

La, Ce, Pr, and Nd chlorides affected the blood picture of rabbits after single or repeated injections. The hemoglobin, leukocyte count,

3,000 to 5,000 mg/kg. Female rats tolerated oral doses of the oxides at 1,000 mg/kg. The chlorides are about half as toxic as the nitrates.

Signs and symptoms of acute poisoning by all the rare earths included ataxia, writhing, slightly labored and depressed respiration, walking on the toes with arched back, and sedation. Death in cats and dogs was due to cardiovascular collapse and respiratory paralysis. Survivors developed generalized peritonitis, adhesions, hemmorrhagic ascitic fluid, true granulomatous peritonitis, and focal hepatic necrosis (Haley 1965).

Intratracheal doses of 50 mg Y_2O_3, Ce_2O_3, and Nd_2O_3 in rats were compared. Ce_2O_3 did not produce serious changes and Nd_2O_3 produced weak to moderate granulomatous and sclerotic granulomatous changes, but Y_2O_3 had produced at 8 mo pronounced granulomatous nodules and emphysematous changes (Stokinger 1981).

B. Chronic Exposure

Stokinger (1981) includes a 12-wk feeding study under his chronic toxicity discussion. At 0.01 to 0.1% in the diet each lanthanide compound tested showed no histological changes in the eight internal organs examined in the necropsy. At 1%, Gd, Tb, Tm, and Yb produced nonspecific liver damage (perinuclear vacuolization and granular cytoplasm).

Guinea pigs daily inhaling mixtures of lanthanides high in fluorides (65% fluorides, 10% oxides, 31% C) at 200 to 300 million particles per cubic foot ($\sim$ 31 to 47 mg/m^3) (1- to 2-μm particles) for 3 yr showed focal hypertrophic emphysema, regional bronchiolar stricturing, and subacute chemical bronchitis but no fibrosis or granulomatosis. A similar experiment with a mixture higher in oxides (39.6% fluorides, 26.4% oxides, and 17% C) revealed dust entrapped within focal atelectatic areas, but again no fibrosis or substantial chronic cellular reaction (Stokinger 1981).

C. Biochemistry

1. Effects on Enzymes

La stimulates succinic dehydrogenase and certain other lanthanides promote the succinic dehydrogenase-cytochrome oxidase system. La and Y inhibit adenosine triphosphate at 0.001 M (Stokinger 1981). YCl$_3$ inhibits mouse liver succinic dehydrogenase (Cochran et al. 1950). In vitro cerium inhibits blood phosphatase but accelerates invertase/saccharase and amylase (Gould 1936). The toxic effects associated with cerium compounds may be an indirect toxic action due to their effect as phosphatase catalysts (Browning 1969). The rare earths are antagonists of thrombokinase (Haley 1965).

2. Metabolism

The lanthanides complex or precipitate proteins. La and nucleic acids form insoluble complexes at physiological pH. By precipitating fibrinogen, the lanthanides have an anticoagulant effect on the blood erythrocyte count, and differential count were all changed (Haley 1965). The lanthanides are blood anticoagulants (Stokinger 1981).

E. Teratogenicity

Ytterbium chloride at intravenous doses of 50 to 100 mg/kg on the 8th day of pregnancy damaged the axial skeleton of the fetus and produced a few nervous system defects and ventral body wall openings (Gale 1975; cited by Shepard 1980). Injections of 8 mg cerium nitrate per egg on the 4th day produced no defects in the chicks (Ridgway and Karnofsky 1952; cited in Shepard 1980). McFee (1964; cited by Browning 1969) observed eye defects in the offspring of rats given [144]Ce (a beta emitter of half-life 282 days) on days 7, 8, or 9 (route not given) of pregnancy.

F. Mutagenicity

No information.

G. Carcinogenicity

Implants of stable Gd and Yb metal pellets in CFW mice appeared to be mild antitumor agents. Certain lanthanide radioisotopes are mild carcinogens (Stokinger 1981).

V. REFERENCES

ACGIH. 1983. TLVs[®] Threshold limit values for chemical substances and physical agents in the work environment with intended changes for 1983-84. Cincinnati, Ohio: American Conference of Governmental Industrial Hygienists.

Browning E. 1969. Toxicity of Industrial Metals. 2nd ed. London: Butterworths.

Chem Eng News 1981. Rare earths. Rhone-Poulenc Chemical Company advertisement. 59(49):14.

Cochran KW, Doull J, Mazur M, DuBois KP. 1950. Acute toxicity of zirconium, columbium, strontium, lanthanum, cesium, tantalum and yttrium. Arch Ind Hyg Occup Med 1:637-650.

Gould BS. 1936. Effects of thorium, zirconium, titanium and cerium on enzyme action. Proc Soc Exp Biol Med 34:381-385.

Haley TJ. 1965. Pharmacology and toxicology of the rare earth elements. J Pharm Sci 54:663-670.

OSHA. 1981. Occupational Health and Safety Administration. Occupational safety and health standards. Subpart 2--Toxic and hazardous substances. Code of federal regulations 29 (Part 1910.1000): 673-679.

Shepard TH.   1980.   Catalog of teratogenic agents, 3rd ed.   Baltimore: The Johns Hopkins University Press.

Stokinger HE.   1981.   Chapter 29.   The metals.   In: Patty's industrial hygiene and toxicology.   3rd ed.   Volume 2A, Toxicology.   Clayton GD, Claytor FE, eds.   New York: A Wiley-Interscience Publication.   John Wiley and Sons, pp. 1493-2060.

ZINC

MAMMALIAN TOXICITY SUMMARY

I. INTRODUCTION

Estimates of the concentration of zinc in the earth's crust range from 5 to 200 ppm. Sphalerite (ZnS) and smithsonite ($ZnCO_3$ or $ZnCO_3$ + Zn_2SiO_4) are the major zinc minerals. Zinc is recovered from zinc ores, copper ores, and mixed ores (Pb-Zn-Cu or Pb-Zn) The extractable zinc concentrations in natural freshwaters are generally 0.03 to 0.06 mg/L or less with values up to 125 mg Zn/L in geographic regions of natural zinc mineralization (Taylor 1982). The amount of zinc in drinking waters, however, also depends on the materials used for piping.

Major uses of zinc include zinc coatings to protect iron and steel, die casting alloys, and brass. Zinc is also used alone (about 10% of total world production) in dry batteries, construction materials, and printing processes (Taylor 1982). Nriagu (1979; cited in Taylor 1982) estimated the relative contributions of anthropogenic sources of atmospheric zinc (the major route for zinc back into the environment): primary zinc production, 32%; wood combustion, 24%; waste incineration, 12%; and iron and steel production, 11%. Fossil fuel burning is not thought to be a major environmental source of zinc (Taylor 1982).

Zinc, atomic number 30, melts at 419.58°C and boils at 907°C. ZnO, ZnS, and $Zn(CN)_2$ are practically insoluble in water (Stokinger 1981).

II. EXPOSURE AND EXPOSURE LIMITS

A. Oral

The present U.S. Environmental Protection Agency and World Health Organization standard for zinc in drinking water is 5 mg/L based on the bitter taste zinc imparts to water at that level. No criterion based on health effects has been established.

Zinc may be present in drinking water usually at concentrations much less than 5 mg/L although softwaters may have such concentrations from materials used in the distribution systems and household plumbing (USEPA 1980). From a study of Boston tap water, levels of zinc were found up to 1,625 µg/L with a mean of 223 µg/L. There was an increase in zinc levels at the tap over finished water levels indicating zinc pickup from the pipes; this was attributed to the soft acidic nature of Boston's water. In the more acidic water of Seattle, zinc levels in tap water were higher than in the finished water 95% of the time with 10% exceeding the 5 mg/L drinking water standard. The highest reported level was 5.46 mg/L (Craun and McCabe 1975, cited by NAS 1977).

Meats, fish, and poultry contain an average of 24.5 mg Zn/kg. Grains and potatoes contain 8 and 6 mg Zn/kg, respectively. Daily intakes via food for the U.S. adult population is 10 to 20 mg Zn/day (EPA 1980). A mean of 0.6 mg Zn/L in United States drinking water would contribute 0.5 to 1 mg Zn/day* (Snyder et al. 1975).

B. Inhalation

Threshold limit values (TLVs) recommended by the American Conference of Governmental Industrial Hygienists (ACGIH 1983) for zinc in workroom air are:

	TWA,[a] mg/m^3	STEL,[b] mg/m^3
Zinc chloride fume	1	2
Zinc oxide fume	5	10
Zinc stearate	(Limits for nuisance particulates)	20

---
[a] TWA = 8-h time-weighted average.
[b] STEL = short-term exposure limit.

OSHA has adopted the fume TLVs (OSHA 1981).

C. Skin

Zinc oxide is a common constituent of many topical preparations (Stokinger 1981).

D. Total Body Burden and Balance Information

The zinc body burden of adult man is estimated at 1.4 to 3 g. Rats fed less than 500 mg Zn/kg did not accumulate zinc, but liver stores increased at ≥ 1,000 mg/kg. Depletion was rapid when the rats were fed a zinc-deficient diet. The whole body half-life of zinc has been estimated to be 933 days (Taylor 1982).

Snyder et al. (1975) give the following zinc balance information for 70-kg Reference Man:

Intake, mg/day		Losses, mg/day	
Food and fluids	13	Urine	0.5
Airborne	< 0.1	Feces	11
		Sweat	0.78
		Hair, nails	0.03
		Menstrual	1

---
* Taylor (1982) cites Craun and McCabe (1975) as finding an average of 0.194 Zn/L in a national drinking water survey.

The National Research Council (National Academy of Sciences 1974) published Recommended Daily Allowances (RDAs) of 15, 20, and 25 mg Zn/day for adults, pregnant women, and lactating women. Surveys of adults of different ages and physiological states including pregnancy and lactation found that normal daily zinc intakes range from 8.6 to 11.4 mg/day or 46 to 67% of the RDAs (Solomons 1982).

III. TOXICOKINETICS

A. Absorption

Zinc bioavailability from food ranges from 10 to 40% (Solomons 1982). Snyder et al. (1975) considered 35% absorption from food zinc to be the best value. Absorption is from the small intestine. Distal portions of the small intestine are involved in absorption of zinc released from a food matrix by digestion (Solomons 1982). Age and content of zinc, protein, phytic acid, and calcium fiber in the diet affect the uptake. For example, zinc-deficient rats showed $\sim$ 98% absorption of zinc in the diet. Plant zinc is less available for human absorption than zinc from animal sources, apparently due to zinc's complexing with phytate (inositol hexaphosphate) and other plant constituents (Taylor 1982).

B. Distribution

1. General

Skin contains up to 20% of whole body zinc. Fresh weight tissue concentrations in muscle, liver, and kidney approximate 55 mg Zn/kg and in prostate, $\leq$ 850 mg/kg. In semen and sperm, concentrations attain 2,000 to 3,000 mg/kg (Taylor 1982).

High zinc concentrations are found in the male reproductive system (highest in the prostate), muscle, bone, liver, kidney, pancreas, thyroid, and some other endocrine glands. The zinc content in the kidney and liver depends on the cadmium concentrations. Zinc is stored in metallothionein and as an essential component of enzymes (USEPA 1980).

2. Blood

Zinc is present in the erythrocytes as the zinc metalloenzyme carbonic anhydrase; in the leukocytes, in several zinc metalloenzymes; in the plasma, bound mainly to albumin; and in the serum, bound to α-macroglobulin and to amino acids (USEPA 1980).

A 1978 National Research Council report compiled extensive data on zinc concentrations in blood, urine, and tissues. That report concluded that 1 mg Zn/L is the mean serum concentration in both men and women. Whole blood has about five times higher concentrations. Red cells contain about 10 times higher concentrations than that in the serum. Serum zinc is lower in women taking oral contraceptives, in pregnant women, and in persons undergoing certain stresses such as infections (USEPA 1980).

Patients given 135 mg Zn/day for 18 weeks to promote healing of leg ulcers showed increased serum zinc levels only if initial serum zinc concentrations were > 1.1 mg/L. Thus, patients with an average of 0.95 mg Zn/L showed an increase to 1.57 mg Zn/L after 6 weeks with no further increase whereas patients with > 1.1 mg Zn/L showed no serum increases in zinc for the 18 weeks (Hallbook and Lanney 1972; cited by USEPA 1980). NIOSH (1975) does not give any information on blood zinc levels in chronically exposed workers and does not recommend monitoring the blood.

Although zinc values in blood serum are generally reliable, the risk of gross error introduction due to sample contamination should not be neglected according to Versieck (1984). Glass, lead, and rubber piping; water; and even high-grade analytical chemicals may introduce zinc contamination.

C. Excretion

Zinc nutritional status affects the long-term biological half-life of zinc, which ranges from $\sim$ 200 to $\sim$ 400 days.

Excretion is largely in the feces. Part of the zinc excreted in the bile to the intestine is reabsorbed. During extreme heat or exercise, much zinc is lost in the sweat. Hair, milk, desquamation of skin, and menstrual blood are other excretory routes (USEPA 1980).

IV. EFFECTS

A. Acute

Zinc ion is poorly absorbed, but salts of strong mineral acids are corrosive to skin and gastrointestinal tract (Stokinger 1981). Ingestion of 2 g or more of zinc produces toxic symptoms in humans. Zinc sulfate (an emetic drug) in these amounts irritates the gastrointestinal tract and causes vomiting (National Academy of Sciences 1980).

Acidic beverages made in galvanized containers have produced mass poisonings. Fever, nausea, vomiting, stomach cramps, and diarrhea develop 3 to 12 h after ingestion (Stokinger 1981).

The taste threshold for zinc in water is $\sim$ 15 ppm. Levels as high as 675 to 2,280 ppm in water are emetic (Stokinger 1981). In laboratory animals, inorganic zinc salts are highly toxic by parenteral routes, but are relatively nontoxic when taken by mouth with rat oral LD_{50}'s in the 1,000 to 3,000 mg/kg range. The organic salts apparently have similar toxicities.

Cows ingesting 140 g Zn/day in contaminated dairy nuts (2% Zn) for a couple of days produced enteritis so severe that on one farm, 7 of 40 affected cows died or had to be slaughtered. Severe pulmonary emphysema and changes in myocardium, kidneys, and liver were observed (Allen 1968; cited by USEPA 1980).

In humans, the lowest toxic concentration for inhaling $ZnCl_2$ dust is 4,800 mg/m³ for 0.5 h.

Zinc metal fume fever has not been successfully produced in animals. "Brass foundryman's ague" and "oxide chills" are similar occupational syndromes due to zinc inhalation. Signs and symptoms appearing 4 to 12 h after exposure to freshly formed fumes include a metallic taste in the mouth or altered tastes of familiar substances, dryness and irritation of the throat, coughing, difficulty in breathing, and chest pain, weakness, fatigue, muscle and joint pain, general malaise, and fever alternating with chills. Recovery occurs within 24 to 48 h. Tolerance is developed quickly but also lost quickly.

B. Chronic

  1. Oral

Patients taking zinc in amounts 10 times the RDA for months and years have not shown any adverse reactions. However, excessive zinc intakes may inhibit copper absorption and thus aggravate marginal copper deficiency (National Academy of Sciences 1980).

Feeding studies in laboratory mammals reveal low oral toxicity. Stokinger (1981) reviewed several "chronic" studies but gave insufficient details for comparison of the relative toxicities. Descriptions of chronic oral studies (> 13 weeks) are also mixed with those of studies as short as 10 days. Toxic signs seen in the animals exposed to the higher amounts (e.g., 1,000 mg ZnO/day in cats and dogs) included glycosuria in dogs; pancreatic fibrous degeneration in cats; and anemia, osteoporosis, and reduced reproductive capacity in rats. Mice given 500 ppm zinc in their drinking water for up to 14 mo exhibited hypertrophy of the adrenal cortex and changes indicating hyperactivity in the pancreatic islets and pituitary gland (USEPA 1980).

In pigs fed diets containing 500 to 8,000 ppm Zn, weight gain and feed intake were reduced at concentrations > 1,000 ppm (EPA 1980). (Deaths occurred as early as 2 weeks after being given a diet containing ≥ 2,000 ppm zinc.)

Lameness and joint afflictions reported in foals living near lead-zinc smelters were related to the zinc exposure. Foals given a diet containing 5,400 ppm Zn showed bone changes, especially in the epiphyseal areas of long bones, and eventually had trouble standing and walking (Willoughby et al. 1972; cited in USEPA 1980).

  2. Inhalation

Long-term inhalation exposure in workers, besides occasional bouts of metal fume fever, produces gastrointestinal disturbances and clinically latent liver dysfunction (Stokinger 1981).

C. Biochemistry

1. Effects on Enzymes

Zinc is essential for the activity of the following enzymes: alcohol dehydrogenase, carboxypeptidase, leucine aminopeptidase, alkaline phosphatase, carbonic anhydrase, RNA-polymerase, and DNA-polymerase. Thymidinekinase may also be zinc dependent (USEPA 1980).

2. Metabolism

Although its biological role has not been completely elucidated, metallothionein is very important in zinc regulation in mammals. Under stressful conditions, its synthesis, zinc accumulation in the liver, and plasma zinc level increase (USEPA 1980). Zinc's involvement in Vitamin A metabolism is discussed by Stokinger (1981). Zinc is ubiquitous in mammalian tissues and is involved in the activity of many enzymes. Thus, zinc participates in many metabolic processes. Excessive zinc intake has been associated with a copper-deficiency anemia. Apparently, zinc interferes with copper and iron metabolism (USEPA 1980).

Interactions of zinc with cadmium, calcium, iron, lead, and drugs are reviewed by USEPA (1980).

3. Synergisms and Antagonisms

Increased zinc uptake or release from bone is accompanied by similar uptake or release of calcium. High zinc-low calcium diets cause severe growth depression in rats.

High zinc-high lead diets cause severe anemia, with zinc reducing or blocking urine ALA excretion* and markedly reducing lead in blood, liver, kidney, and bone.

Zinc administered simultaneously with cadmium to pregnant hamsters (but not 12 h after) prevented malformations caused by cadmium alone. Zinc appears to reduce the toxic effects of cadmium. By inducing the production of metallothionein, zinc protects against the teratogenic action of mercury.

Zinc has been positively correlated wtih lithium in drinking water and arteriosclerotic heart disease mortality studies.

The toxic symptoms of rats exposed to sporidesmin (the major toxin in the fungus Pithomyces chartarum, which causes liver toxicity in New Zealand livestock) were prevented or reduced by oral treatment with zinc salts (Taylor 1982).

---

* Lead suppresses ALA-D activity, but zinc counteracts the effect.

4. Physiological Requirements

Zinc appears to be essential in enzymes and enzymatic functions; DNA, RNA, and protein synthesis; carbohydrate metabolism; utilization of nitrogen and sulfur; and cell division and growth. It is essential for spermatogenesis, ova formation, and fetal nutrition. It also appears to be involved in pituitary gland and adrenal gland metabolism. Absolute amounts required range from 1.1 mg/day in infancy, 2.2 mg/day in childhood and adulthood, 2.8 mg/day at puberty, 2.55 to 3.0 mg/day in pregnancy, and 5.45 mg/kg during lactation (Taylor 1982).

Zinc deficiency signs and symptoms in animals and humans include loss of appetite, decreased sense of taste, failure to grow, impaired wound healing, and other skin changes. Severely deficient humans exhibit hypogonadism, dwarfism, acrodermatitis enteropathica, and malabsorption syndrome. Marginal zinc deficiencies in the United States have been reported (National Academy of Sciences 1980).

D. Specific Organs and Symptoms

1. Lung

Acute pulmonary symptoms in metal fume fever were described above. Pulmonary lactic acid dehydrogenase (LDH) isoenzyme factor 3 may be transiently elevated.

A fatal case of extensive lung fibrosis has been described in Italy in a worker who had been exposed to zinc stearate for 29 years in a rubber factory. No adverse effects due to zinc stearate exposure have been reported in the United States (Stokinger 1982).

2. Blood

Polymorphonuclear leukocytosis occurs during acute bouts of metal fume fever. The white count rises to 20,000 cells/mm^3 (Stokinger 1982).

3. Gastrointestinal Tract

Pressure in the stomach region, nausea, and weakness, symptoms suggestive of gastric or duodenal ulcers, have occurred in zinc workers and workers exposed to zinc during welding and cutting of galvanized iron and steel (Stokinger 1982).

4. Skin

Zinc oxide exposure combined with lack of personal hygiene contributed to an occupational dermatitis caused by blocked sebaceous glands. Generally, zinc oxide has a low potential for skin irritation. It is an ingredient of many topical dermatological preparations (Stokinger 1982).

E. Teratogenicity

No information was found by USEPA (1980) that excessive zinc causes developmental malformation, but zinc deficiencies may. Negative results are described by Stokinger (1981).   Shepard (1980) cited a 1977 study by Chang et al. in which the offspring of mice given 20 mg/kg i.p. on days 8, 9, 10, or 11 of gestation showed a delay in ossification and some skeletal malformations.

F. Mutagenicity

USEPA (1980) found no information suggesting that zinc is a mammalian mutagen. Stokinger (1981) discusses some negative results.

G. Carcinogenicity

Injection of zinc salts into the testes of laboratory animals has induced testicular tumors, but zinc is not otherwise directly associated with oncogenicity. Zinc appears to be indirectly involved, however, since zinc is required for tumor growth (although initially excessive zinc gives some protection). Zinc deficiency does not appear to have an etiological role in human cancer development, but neoplastic tissues may have lower zinc levels (USEPA 1980).

V. REFERENCES

ACGIH. 1983. TLVs® threshold limit values for chemical substances and physical agents in the work environment with intended changes for 1983-84. Cincinnati, OH: American Conference of Governmental Industrial Hygienists.

National Academy of Sciences. 1974. Recommended dietary allowances. 8th ed. Washington, DC: Printing and Publishing Office, National Academy of Sciences.

National Academy of Sciences. 1980. Recommended dietary allowances. 9th ed. Washington, DC: Pringing and Publishing Office, National Academy of Sciences.

NIOSH. 1975. National Occupational Safety and Health Administration. Criteria for a recommended standard: Occupational exposure to zinc oxide. Washington, DC: U.S. Government Printing Office. HEW Publication No. (NIOSH) 76-104.

NRC. 1978. National Research Council. Zinc. Baltimore: University Park Press.

OSHA. 1981. Occupational Safety and Health Standards. Subpart 2--Toxic and hazardous substances. Code of federal regulations 29 (Part 1910.1000): 673-679.

Shephard TH. 1980. Catalog of teratogenic agents, 3rd. Baltimore: The Johns Hopkins University Press.

Snyder WS, Cook MJ, Nasset ES, Karhausen LR, Howells GP, Tipton IH. 1975.
International Commission on Radiological Protection. Report of the task
group on reference man. New York. ICRP Publication 23.

Solomons NW. 1982. Biological availability of zinc in humans. Am J Clin
Nutr 35:1048-1075.

Stokinger HE. 1981. Chapter 29. The metals. In: Patty's industrial
hygiene and toxicology. 3rd ed. Volume 2A. Clayton GD, Clayton FE, eds.
New York: A Wiley-Interscience Publication.   John Wiley and Sons,
pp. 1493-2060.

Taylor MC, DeMayo A, Taylor KW. 1982. Effects of zinc on humans, laboratory
and farm animals, terrestrial plants, and freshwater aquatic life. Crit Rev
Environ Control 12(2):113-181.

USEPA. 1980. U.S. Environmental Protection Agency. Ambient water quality
criteria for zinc. Springfield, VA: National Technical Information Service.
PB81-117897.

ZIRCONIUM

MAMMALIAN TOXICITY SUMMARY

I. INTRODUCTION

The toxicity of zirconium was briefly reviewed by Stokinger (1981). However, his references were no more recent than those of the extensive review by Smith and Carson (1978).

A. Occurrence and Production

The crustal abundance of zirconium has been variously estimated at 160 to 280 ppm. Zirconium is a constituent of numerous rocks including granular limestone and granite. The silicate zircon is its most abundant and widely distributed mineral; baddeleyite, $4[ZrO_2]$, is the next most important mineral. In the United States, zirconium is recovered as a by-product of titanium production from beach sands. Numerous states have zircon deposits. Zirconium also appears in the wastes from processing bauxite and phosphate rock. The major purification process is the Kroll process, in which $ZrCl_4$ is reduced by Mg (Smith and Carson 1978).

B. Uses

Major uses for zircon are foundry sands and refractories. Zirconates are used in electrical ceramic compositions. The largest use for the metal is in Zircaloy alloys as containers for fuel rods and as pressure tubes in water-cooled nuclear reactors and Navy nuclear submarine propulsion systems. Other metal uses are as a "getter" for residual nitrogen and oxygen in vacuum and other electronic tubes, chemical apparatus, minor Zr-based alloys, and a grain refiner in magnesium alloys. Ferrozirconium and ferrosilicon zirconium are used as deoxidizers and scavengers in the steel industry.

Zirconium naphthenate and other soaps have been used as substitutes for lead and other heavy metal driers in paint formulations. The major use of water-soluble zirconium compounds (e.g., the sulfate, oxychloride, and oxynitrate) is in tanning. Other uses are in textile finishing to impart algae and mildew resistance and water repellency. Zirconium compounds are used in over-the-counter poison ivy remedies. Acid-soluble zirconium complexes were used in underarm deodorants; their use in aerosols was banned in 1974.

C. Chemistry

Zr^{4+} ions exist only in strongly acidic solutions at very low (< 9.1 ppm) zirconium concentrations. Although the names of its salts frequently incorporate the word zirconyl, this ion ($Zr:O^{2+}$) is not present in aqueous solution or in the solid salts. In aqueous solution, salts hydrolyze to give polymeric hydroxy zirconium species and release the mineral acids with strong lowering of the pH. In fact, this pH lowering is the only environmental effect to be expected since the hydrous zirconia formed is very insoluble (Smith and Carson 1978).

II. EXPOSURE AND EXPOSURE LIMITS

A. Oral

It has been speculated that zirconium contamination of superphosphate fertilizers (146 ppm in only one report) could represent a route into human foods. However, plants, generally do not translocate zirconium above the roots so only root crops would be affected. Contaminated algae, algin and alginates, and shellfish were also considered possible sources of zirconium into human food. Levels of up to 6.64 ppm have been reported in meat, poultry, eggs, and dairy products (determination was by activation analysis). No drinking water concentrations were found, but surface fresh waters contain up to 0.02 mg/L. Some toothpastes and dental prophylactic preparations have contained Zr compounds. Insoluble zirconium compounds in lipsticks might also contribute to oral intake (Smith and Carson 1978).

B. Inhalation

Few data were found on zirconium in ambient air, but it should be expected to be present because of its high natural background concentration in soils. Little or no hazard would be expected from industrial emissions of insoluble zircon and ZrO_2 (Smith and Carson 1978).

The threshold limit value as a time-weighted average in workroom air for an 8-h day is 5 mg Zr/m^3. For zirconium compounds the short-term exposure limit is 10 mg Zr/m^3/15 min (ACGIH 1983). The OSHA permissible exposure limit is 5 mg Zr/m^3 (OSHA 1981).

C. Dermal

Nonaerosol deodorants, lipsticks, and poison ivy lotions allow dermal contact of zirconium compounds (Smith and Carson 1978).

D. Total Body Burden and Balance Information

Using rodent data of Schroeder and Balassa and later data of Schroeder, Smith and Carson (1978) estimated that the human body burden of a 70-kg human would be 308 mg; however, since the atomic absorption data used may be questionable, they also calculated a body burden of < 7 mg and a dietary intake of 0.053 mg Zr/day from the data of Hamilton and Minski (1972/73; cited by Smith and Carson 1978). Still other data would indicate that food of plant origin should contribute ~ 1 mg Zr/day (Smith and Carson 1978).

According to Snyder et al. (1975) the zirconium balance for 70-kg reference man (mg/day) is: intake from food and fluids, 4.2; losses from urine, 0.15, and feces, 4.

III. TOXICOKINETICS

A. Absorption

Data on gastrointestinal absorption are generally based on experiments with ^{95}Zr (radiozirconium) and range from 0 to 2%. Atomic absorption data on urinary excretion of zirconium by two men would indicate $\geq$ 20 to 33% absorption. Absorption of soluble compounds is better from the lungs. In guinea pigs exposed nose only to ^{95}Zr oxalate in air for 0.5 h, 12.10% of the total dose received was in the blood within that time and 17.8% was in other internal organs (Smith and Carson 1978).

B. Distribution

1. General

Zirconium was found in lungs, liver, and bone of rats fed zirconyl chloride and in liver, kidney, heart, lung, and spleen at concentrations up to 46 ppm wet weight in mice fed 2.66 ppm Zr in the diet and 5 ppm in the drinking water for life (Smith and Carson 1978).

2. Blood

Average whole blood concentrations compiled by Smith and Carson (1978) were ~ 0.01 to 0.02 ppm and ~ 2 to 3 mg% ash weight. Serum averages were 0.11 and 0.68 ppm. Erythrocytes contain 6.18 ppm. Only 0.001 ppm was found in the mineral part of plasma.

3. Adipose

One report gave human fat content at 18.71 (range 2.94-42.27) ppm. The atomic absorption data may be too high. No other fat concentrations were found by Smith and Carson (1978) in their extensive review.

C. Excretion

Dietary zirconium is excreted in both urine and feces with either predominating. After intentional administration of zirconium compounds, fecal excretion predominates for several days due to low absorption and hepatic excretion. Soluble Zr compounds in the lungs may be removed by the pulmonary lymph nodes. They may also pass directly from the lung parenchyma to the blood. Small insoluble particles, escaping mucociliary clearance, have biological half-lives in humans of > 200 days (Smith and Carson 1978).

IV. EFFECTS

A. Acute and Other Short-Term Exposures

Zr citrate and Zr malate were studied in the nuclear industry as a potential treatment of accidental contamination by numerous long-lived radio-elements. Two persons given Zr malate in 50-mg i.v. injections suffered vertigo and labarinthitis. Human workers have generally experienced no acute toxicity in the workplace.

The acute oral toxicity of zirconium compounds is generally very low if other compound moieties are nontoxic and if the compound does not hydrolyze to produce a strong acid. Reported oral LD_{50}'s range from 61 mg Zr/kg mouse for $InZrF_7$ to 2,290 mg Zr/kg rat for Na Zr sulfate. Acute oral toxic doses produce gastrointestinal tract necrosis, renal and hepatic dystrophy, and hemorrhage. Weakness and death by respiratory and circulatory paralysis occur. Subacute oral doses (short-term repeated dosing) cause weight loss, weakness, diarrhea, liver damage, and death (Smith and Carson 1978).

Given by s.c. injection, repeated doses of 20 to 65 mg Zr/kg as $ZrCl_4$ produced a temporary change in the conditioned reflexes of rats for several days after the injection (Smith and Carson 1978).

B. Chronic Exposure

Few chronic studies have been done. Dogs fed 100 mg Zr citrate per kg every 1 to 4 days for ~ 5 mo did not show any toxic effects. Mice given 5 ppm Zr as the sulfate in drinking water and 2.66 ppm in the diet from weaning until natural death showed a slightly reduced survival rate and a lower weight gain than control mice who received Zr only in the diet. Survival or longevity was not affected in a similar test with rats and the weight effect was not consistent.

In one study purporting to show the protective effect of Zr citrate to reduce the incidence of bone cancer from ^{90}Zr deposition, i.v. Zr treatments increased the death rate before any tumors appeared (147 days) from 4 to 6.6 times compared with controls given only ^{90}Sr.

Stick deodorants and poison-ivy remedies containing zirconium compounds have produced skin epithelioid* granulomas in humans as a delayed hypersensitivity reaction. The reaction could not be duplicated in laboratory mammals.

Lung changes described in zirconium workers might have been due to other constituents of the dusts. Fibrous lung changes, but not necessarily fibrosis, and liver necrosis were observed in animals dosed intratracheally with various zirconium compounds. Lung granulomas from lung intake (by rabbits) were reported only once (Smith and Carson 1978).

C. Biochemistry

1. Effects on enzymes

Zirconium salts inhibit many enzymes such as amylase, emulsin, adenosine triphosphatase, yeast invertase, blood phosphatase, glutamic-

---

*   An epithelioid granuloma appears to synthesize a granuloma-perpetuating factor; no foreign body can be detected.

oxalacetic and glutamic-pyruvic transaminases, glutaminase, and pyrophosphatase
The initial inhibitions of ATPase and pyrophosphatase are followed by prolonged
activation (Smith and Carson 1978).

2. Metabolism

Little information was found regarding biological binding or chemical changes of zirconium in the body (Smith and Carson 1978). Zirconium may
bind the phosphoric acid of nucleic acids.

Zirconium deposits in the trabecular bone probably bound to proteins. Colloidal zirconium particles are generally phagocytized and
accumulate in the reticuloendothelial system.

3. Antagonisms and synergisms

No information.

D. Specific Organs and Systems

1. Lungs

Fibrous lung changes occur in animals with loss of respiratory
alveolar tissue and occasionally epithelial hyperplasia (Smith and
Carson 1978).

2. Skin

Skin granulomas have occurred in hypersensitive people using
zirconium-containing deodorants and poison-ivy remedies (Smith and
Carson 1978).

3. Liver and kidney

Acutely poisoned animals exhibited dystrophic or necrotic lesions
of the liver and kidney (Smith and Carson 1978).

E. Teratogenicity

No information.

F. Mutagenicity

No information.

G. Carcinogenicity

No adequate tests of carcinogenicity were found.

V. REFERENCES

ACGIH. 1983. TLVs® Threshold limit values for chemical substances and physical agents in the work environment with intended changes for 1983-84. Cincinnati, Ohio: American Conference of Governmental Industrial Hygienists.

OSHA. 1981. Occupational Safety and Health Admin. Occupational safety and health standards. Subpart 2--Toxic and hazardous substances. Code of federal regulations 29 (Part 1910.1000) pp. 673-679.

Smith IC, Carson BL. 1978. Trace Metals in the Environment. Volume 3- Zirconium. Ann Arbor, MI: Ann Arbor Science Publishers, 405 pp.

Stokinger HE. 1981. Chapter 29. The metals. In: Patty's industrial hygiene and toxicology, 3rd ed., volume IIA, Toxicology. Clayton GD, Claytor FE, eds. New York: Wiley-Interscience, John Wiley & Sons, pp. 1493-2060.